HOME-GROWN
MUSHROOMS
FROM SCRATCH

Home-Grown Mushrooms from Scratch: *A Practical Guide to Cultivating Mushrooms Outside and Indoors*
Copyright © 2018 by Magdalena Wurth and Herbert Wurth

Originally published in Germany as *Pilze selbst anbauen* by Löwenzahn in der Studienverlag GesmbH in 2015 and in the UK as *Home-Grown Mushrooms from Scratch* by Filbert Press in 2017. First published in North America in revised form as *How to Grow Mushrooms from Scratch* by The Experiment, LLC, in 2018. This paperback edition first published in 2025.

All rights reserved. Except for brief passages quoted in newspaper, magazine, radio, television, or online reviews, no portion of this book may be reproduced, distributed, or transmitted in any form or by any means, electronic or mechanical, including photocopying, recording, or information storage or retrieval system, without the prior written permission of the publisher.

The Experiment, LLC
220 East 23rd Street, Suite 600
New York, NY 10010-4658
theexperimentpublishing.com

This book contains the opinions and ideas of its author. It is intended to provide helpful and informative material on the subjects addressed in the book. It is sold with the understanding that the author and publisher are not engaged in rendering medical, health, or any other kind of personal professional services in the book. The author and publisher specifically disclaim all responsibility for any liability, loss, or risk—personal or otherwise—that is incurred as a consequence, directly or indirectly, of the use and application of any of the contents of this book.

THE EXPERIMENT and its colophon are registered trademarks of The Experiment, LLC. Many of the designations used by manufacturers and sellers to distinguish their products are claimed as trademarks. Where those designations appear in this book and The Experiment was aware of a trademark claim, the designations have been capitalized.

The Experiment's books are available at special discounts when purchased in bulk for premiums and sales promotions as well as for fundraising or educational use. For details, contact us at info@theexperimentpublishing.com.

The Library of Congress has cataloged the earlier edition as follows:

Names: Wurth, Magdalena, author. | Wurth, Herbert, author.
Title: How to grow mushrooms from scratch : a practical guide to cultivating
  portobellos, shiitakes, truffles, and other edible mushrooms / Magdalena
  Wurth and Herbert Wurth.
Description: English language edition. | New York, NY : The Experiment, LLC,
  2018. | "Originally published in Germany as Pilze selbst anbauen by
  Löwenzahn in der Studienverlag GesmbH in 2015 and in the UK as Home-Grown
  Mushrooms from Scratch by Filbert Press in 2017." | Includes
  bibliographical references and index.
Identifiers: LCCN 2018014463 (print) | LCCN 2018023009 (ebook) | ISBN
  9781615195107 (Ebook) | ISBN 9781615194919 (cloth)
Subjects: LCSH: Mushroom culture. | Mushrooms.
Classification: LCC SB353 (ebook) | LCC SB353 .W937 2018 (print) | DDC
  635/.8--dc23
LC record available at https://lccn.loc.gov/2018014463

ISBN 979-8-89303-097-6
Ebook ISBN 978-1-61519-510-7

Cover design by Beth Bugler
Text design by Sophie Appel

Author photograph by Benedikt Wurth
Cover and interior photographs by Magdalena Wurth unless otherwise indicated
Cover photograph of mushrooms in crates by Walter Haidvogl
Illustrations by Anna Folie

Manufactured in India

First paperback printing July 2025
10 9 8 7 6 5 4 3 2 1

HOME-GROWN
MUSHROOMS
FROM SCRATCH

A Practical Guide to Cultivating Mushrooms Outside and Indoors

MAGDALENA WURTH *and* **HERBERT WURTH**

THE EXPERIMENT

NEW YORK

Contents

Authors' Note	6
Introduction: The World of Mushrooms	8

1. Cultivating Mushrooms in the Garden — 14

Mushrooms from Logs	18
Inoculation with Mushroom Spawn	19
The Development Stages of the Mushroom Garden	25
Cultivated Mushrooms as Features of Garden Design	32
Mushrooms from Straw	35
Inoculating with King Stropharia Spawn	37
Cultivating Oyster Mushrooms on Straw	40
Mushroom Portraits	42

2. Cultivating Mushrooms Indoors — 56

Mushrooms on Straw Pellets—Practical and High Yielding	56
Mushroom Kits—The Perfect Way to Cultivate in Wintertime	58
Mushrooms from Compost	61
Additional Mushroom Portraits	64

3. Cultivating Mushrooms in Woodlands and Fields — 67

Cultivating Truffles—The Secrets of Mycorrhizal Fungi	67
Important Points in Truffle Cultivation	68
Establishing a Bed for Mushrooms	70
Preparing a Mushroom Bed	71
Speciality: Shaggy Mane in Beds and Containers	72
Inoculating Tree Stumps—Edible Mushrooms Encouraging Biological Succession	75

4. Container-Grown Mushrooms for Courtyards, Balconies, and Patios — 80

Setting Up a Mushroom Garden on the Balcony or Patio	80

5. "Protected" Environments for All Seasons — 85

Erecting a Protective Roof in the Garden	85
Small Indoor Greenhouses for Edible Mushrooms in Winter	86
A Mushroom House for the Garden	86
Special Technique: Reishi in Pots	87

6. Pests and Competing Organisms in Mushroom Cultivation	**90**
7. Propagating Mushrooms—From Spore to Spawn	**96**
For Specialists, Experienced Growers, and Novice Experimenters	96
An Overview of the Working Stages of Mushroom Propagation in the Lab	101
8. The Use of Mushrooms in Medicine	**108**
9. Recipes and Processing Edible Mushrooms	**118**
Shiitake	118
Oyster Mushrooms	122
Sheathed Woodtuft	124
Button Mushrooms (Champignons)	125
King Oyster Mushroom	126
Lion's Mane	128
Wood Ear (Jelly Ear)	128
King Stropharia (Garden Giant)	129
Making Mushroom Powder and Processing Medicinal Mushrooms	130
10. Marketing Organic Mushrooms on a Small Scale	**133**
Case Study: Home Produced Fresh Mushrooms at Gasthaus Seidl, Vienna, Austria	133
Appendixes	136
Bibliography and Recommended Reading	139
Index	140
Acknowledgments	144
About the Translator	144

Authors' Note

My Fascination with Mushrooms

For me, mushrooms are associated with a childlike fascination. Each mushroom has its own unique appearance, a specific *modus vivendi*, and unmistakable flavor. Whenever I enter the world of mushrooms—whether in a woodland or in my own garden—I experience a deep sense of joy. I think the easiest way to understand my fascination with these tiny forms is by describing my morning rounds in the garden to see how everything's doing. On close inspection, tiny formations of mushrooms, just hours old, suddenly come into view. I can hardly wait for them to appear, and it's exciting when the work and waiting finally turns into a delicious meal.

For my family (as far back as I can remember), mushrooms have always been of great importance and are just as significant to us as our own lovingly tended vegetable garden. Both contribute color and diversity to our meals. For me, there's nothing more satisfying than having my hands full with gardening tasks such as harvesting vegetables, drying herbs, or preserving my own mushrooms. Our kitchen is always full of people and, along with the garden, is the family's main workplace. Although it often requires a lot of work and attention, I greatly value home-grown food, and the taste of summer captured in aromatic, dried mushrooms in the cold months is priceless.

Another aspect of mushrooms that is important to me is their breeding. To propagate them in one's own lab requires good instincts, patience, and expert knowledge of the needs and life cycles of each mushroom species being bred. I've spent countless hours in the lab with my father, observing his microbiological work and looking on as he propagates mushrooms. After a long time observing and following my father's lead, I realized my own interests also lay in exploring the mystery of mushrooms. This practical book is the result of my father's many years of experience and my own motivation to delve deeper into the world of mushrooms, not to mention the support of the publisher of the original German language edition, Löwenzahn. We are excited to share our knowledge and experience with other interested people.

Magdalena Wurth

The Origins of Our Mushroom Garden

Herbert and Magdalena Wurth

I spent a lot of time in the forest as a child, and even back then, mushroom hunting was a family tradition. Later, as I began my professional life as a chemist, I worked together with collaborators at TU Wien, the Vienna University of Technology, on a fascinating project that involved fungi that break down cellulose for the further processing of agricultural by-products. Because it involved similar methodologies as this project, cultivating mushrooms at home suddenly seemed very accessible. In those early years, commercial mushroom spawn was hard to come by, a situation that turned out to be a blessing in disguise. Over the years we accumulated a lot of experience in the fields of microbiology, mushroom spawn production, and the cultivation of edible mushrooms on myriad substrates. By 1984 we had inoculated 100 straw bales with king stropharia mushroom. More recently we have been focusing our efforts on cultivating and breeding edible mushrooms on logs, and we grow our shiitake mushrooms by emulating the traditional Japanese method.

Twenty-five years ago we relocated and thus had the opportunity to establish a new mushroom garden and gain valuable experience in design. Through our cooperation with Arche Noah, we noticed great interest from gardeners in cultivating edible mushrooms using natural methods. Personally, I'm most interested in the challenges presented by more difficult to cultivate mushrooms like reishi. Working with fungi has allowed for profound insights into the inner secrets of these fascinating organisms.

We have a family-owned business in Austria, the Waldviertler Pilzgarten. We're excited about the continuing development and implementation of new ideas in the field of fungiculture. We hope you will see this book as an opportunity to start growing edible mushrooms in your garden, cellar, or courtyard, on your patio or balcony, or even in the kitchen.

Herbert Wurth

Harvested mushrooms ready to cook and eat

Introduction: The World of Mushrooms

What are Fungi?

Fungi have no chlorophyl within their cells. Unlike most plants, they do not use photosynthesis to produce energy. Fungi have developed a fascinatingly large number of methods of obtaining nutrients, each characteristic of individual species.

Many species are saprophytes, that is, they feed off dead organic matter. Because they can break down components of wood such as cellulose, saprophytic fungi have an important role in the forest as decomposers—they break down dead wood and other dead plant material into simple organic compounds. Humus is thus created. So wood, straw, or compost are often used as growing media when cultivating edible mushrooms.

Saprophytes can be placed into two groups—primary and secondary decomposers:

- The former are fungi that are able to break down an unaltered "raw" growing medium. Examples of this group include oyster mushrooms and sheathed woodtuft.
- Secondary decomposers, by contrast, require that their nutritional basis first be macerated by microorganisms. Button mushrooms and shaggy mane are included in this group.

Another group of fungi live parasitically. These species can be especially problematic in the fields of agriculture and silviculture, as in the case of honey fungus (*Armillaria spp.*), which is common in trees and woody shrubs. Parasitic fungi can also infect living (often already weakened) organisms and sap them of energy and nutrients.

The third group of fungi are the mycorrhizae, including, for example, burgundy truffle (*Tuber aestivum* var. *uncinatum*), porcini, and chanterelles. The word *mycorrhiza* comes from the

Greek words *mykos* (fungus) and *riza* (root). We distinguish between three types:
- Ectotrophic mycorrhizae attach themselves to the surface of the roots of certain higher plants.
- Endotrophic mycorrhizae are widely distributed and have the ability to integrate their hyphae into root and bark cells.
- Ectendomycorrhizae bridge the gap between the two groups described above. Likewise these live in symbiosis with plants.

Plants and fungi both benefit from this partnership. A fine, practically invisible network of fungal hyphae encases the roots of plants and solubilizes nutrients for them. In return, the plants supply the fungi with carbohydrates (sugar). The plant's root system increases in effective size, which increases its capacity to take up nutrients and water. Such partnerships between fungi and plants are complex and in many cases have yet to be researched. This is also the reason for the low success rate in cultivating wild fungi such as porcini and chanterelles. Experiments with mycorrhizal fungi have often pointed to the conclusion that cultivation without a symbiotic partner is impossible. One of the few examples of successful mycorrhiza cultivation are truffles. Incidentally, a symbiotic mycorrhizal partner is also a requirement in orchid breeding.

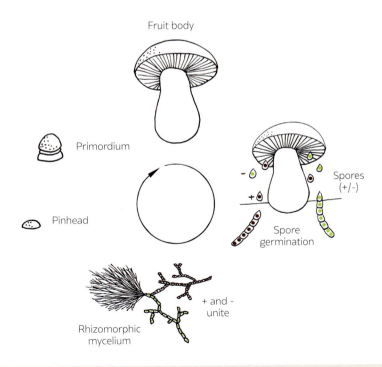

Structure and life cycle

HOME-GROWN MUSHROOMS FROM SCRATCH

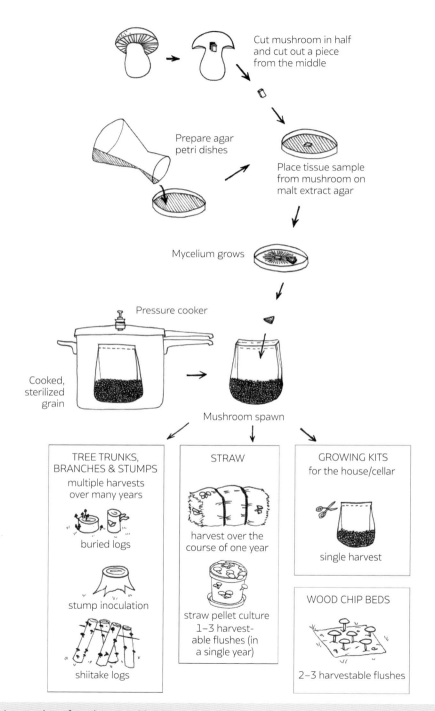

An overview of mushroom cultivation

INTRODUCTION: THE WORLD OF MUSHROOMS

The authors' mushroom garden

Structure and Life Cycle of a Mushroom

When we talk about mushrooms, we mean the above-ground fruit body of the larger fungus organism. Indeed, the entire organism consists of far more than just the familiar cap, gills, and stem of the fruit body; it is mostly the mycelium that grows underground. Mycelium is made up of hyphae, which are somewhat like the fine root hairs on tree roots. Still, the outward appearance of the mushroom is an important distinguishing feature of fungi.

Spores, which are somewhat analogous to the seeds of plants, are produced and found in the gills, ridges, teeth, or pores of edible and medicinal mushrooms. They are microscopically small and, at the appropriate moment of the fungus' development cycle, are disbursed into the surrounding area, often with the help of wind, water, or other organisms. Positively (+) and negatively (-) polarized spores germinate when they meet under favorable environmental conditions in a suitable substrate. Hyphae then proliferate and rhizomorphic mycelium is created. As development continues, dense networks of mycelia form, upon which nodules called primordia develop, from which mushrooms eventually emerge. Mushroom emergence represents the completion of the fungal life cycle, which can now begin anew.

An Overview of Mushroom Propagation

When propagating mushrooms, there are two general methods of propagation that are practiced. One is germinating the spores collected from mushrooms in a lab. The other method is to propagate from a tissue sample of a mushroom in an appropriate medium. Mycelium grows forth from the tissue sample, which can then be further propagated. The initial substrate is typically nutrient agar, as this is ideal for mycelial growth. Then a piece of grown-through agar is transferred to cooked, sterilized grain. Once the fungus colonizes this new substrate, it can be used to inoculate wood, straw, or compost.

OUR TIP: *Mushroom cultivation is a fascinating and expansive topic. The deeper you immerse yourself in it, the easier it will be to understand fungi and have success cultivating them. See the chapter entitled "Propagating Mushrooms" (p. 96) for practical advice and further details.*

The Importance of Fungi in Nature and for Us Humans

Fungi are ubiquitous in nature and also in the day-to-day lives of people. Their spores, for example, move about invisibly in the atmosphere. Nearer to the ground, they are part and parcel of any given forest floor. And one need only consider their extensive medical and economic significance to appreciate how important fungi are to people. In the baking and dairy industries, cultivated strains of yeasts and other genera such as *Penicillium* (in cheese production) are used. Penicillin, which is a metabolic product of the mold fungus *Penicillium chrysogenum*, was a revolutionary discovery for fighting bacterial infections in people and animals. Medicinal mushrooms also have a long history of use in traditional Chinese medicine. On the other hand, fungi can be harmful for people, animals, and plants in that they are capable of infecting living organisms, sometimes with disastrous consequences.

A large part of disease control in plants has to do with fighting and controlling parasitic fungi. Widespread fungal diseases of plants include late blight in potatoes (*Phytophthora infestans*), common bunt of wheat (*Tilletia caries*), botrytis bunch rot of wine grapes, and gray mold of soft fruits and bulb crops (*Botrytis cinerea*).

Fascinating and often invisible is the omnipresent participation of fungus in the decomposition process of organic matter. Fungi and bacteria contribute immensely to homeostasis in nature by breaking down organic compounds into substances available to plants and forming soil. Mycorrhizal fungi optimize the development of many plants in dynamic environments. And let's not forget the delicious wild mushrooms that are to be found in forests and fields. More and more people are captivated by mushroom hunting. A wonderful alternative to pillaging nature's bounty is the cultivation of edible mushrooms.

Hericium coralloides *(coral tooth fungus) in a beech forest*

Oyster mushrooms in various stages of ripeness

A bountiful harvest of shiitakes in the mushroom garden

1. Cultivating Mushrooms in the Garden

There are a huge number of different locations where fungi thrive. When walking through the woods you'll find delicious edible mushrooms again and again. Fortunately, there are more than just mycorrhizal fungi that live in symbiosis with higher plants. When cultivating edible mushrooms, wood, straw, or compost are the most common growing media. The more precisely you are able to observe the processes of nature, the more you can learn from them. These observations can be used to optimize the growing conditions for home-grown mushrooms.

Edible mushrooms will not grow just anywhere and at any time. To start your own mushroom garden, you will need to consider several factors regarding location. Mushrooms prefer a moist microclimate, so your site should be protected from wind and at least semi shady. So what is the best site for your mushroom garden? Deciduous trees, bushes, or a hedge make for ideal shading. The presence of moss and ferns at a given site is a fairly reliable signal that mushrooms would thrive there. Even the shady bank of a brook or edge of a pond could support cultivated mushrooms. Most fungi do comparatively poorly under conifer trees, as it can be too dry there even with heavy rains.

Another important factor for siting your prospective mushroom garden is the availability of water for irrigating your mushroom cultures. Individual mushroom species have different irrigation needs, which we will discuss in detail later in this chapter. Fungi are much more flexible in terms of what kind of soil they need. One need only ensure that drainage is sufficient for the soil not to become waterlogged.

Size and Basic Configuration of a Mushroom Garden

Once you've found an appropriate site for your mushroom garden, questions remain about its size and basic configuration. In order to provide for your own mushrooms as close to year round as possible, grow several different kinds and use several different growing media (logs, straw, wood chips, etc.). Each growing medium has its own set of advantages and disadvantages—consider these when choosing which kind of mushroom to grow.

We recommend for the average needs of a three- to four-person household about six logs (3 feet [1 m] long, 4–6 in [10–15 cm] diameter) for shiitake and three thicker logs (3 feet [1 m] long, 8–14 in [20–35 cm] diameter) for one of the many kinds of oyster mushroom or sheathed woodtuft. Additionally, two straw bales can be inoculated with king stropharia or an oyster mushroom. Logs inoculated with shiitake should be stacked up off the ground. The other logs are to be cut in thirds and buried 4 in (10 cm) into the ground (see Establishing a Mushroom Garden, p. 26). The inoculated straw bales also require contact with the soil in order to produce mushrooms. All in all, you will need about 75 ft^2 [7 m^2] for this assortment of mushrooms in these quantities, which you will be able to harvest from spring through autumn. Additionally, you could inoculate some logs with enokitake—an exquisite winter mushroom.

For the most part, climatic conditions will determine when and how many mushrooms will be harvested. Shiitakes are a bit of an exception, as the grower can also influence when the logs will flush with mushrooms (see Mushrooms from Logs, p. 18). See to it that logs are kept as free of slugs as possible. Slugs are attracted to the scent of tiny emerging mushrooms and love to munch on them (see Pests and Competing Organisms in Mushroom Cultivation, p. 90).

Each fungus needs a certain amount of time to colonize its growing medium, ranging from a few months to up to two years. Some mushrooms, like oyster mushrooms, can be grown on either wood or straw. King stropharia prefers straw and has long been grown because of its high yields and ease of cultivation. Mushrooms typically colonize straw faster than they do wood.

When starting your first mushroom garden, it is only natural to want to see quick results, and here are some ways to achieve this. Buying pre-inoculated logs from a reputable mushroom grower can be a reliable way to quickly achieve your first successful mushroom harvest. Alternatively, since mushrooms quickly colonize straw, use a straw-loving species for your first inoculations. This should usually be done in spring, as autumn inoculations take longer to establish (inoculating with king stropharia in spring will yield mushrooms before summer).

In addition to pre-inoculated logs, procuring mushroom spawn to inoculate logs is another good way to get into cultivating shiitake, oyster mushrooms, and others. Logs may take longer to colonize than straw, but they can produce for up to five years. Because fungi metabolize straw much faster, inoculated straw typically only yields for about a year. Using logs makes for a mushroom garden that persists and sustains.

In later chapters, the advantages and disadvantages of various substrates will be discussed in greater detail.

Mushroom Spawn—The Most Important Component of Mushroom Cultivation

In order to cultivate mushrooms at home, you are going to need the right mushroom spawn. Just like in vegetable gardening, where you can either buy seedlings or grow them yourself from seed, in mushroom cultivation you can either make your own spawn or order it from a supplier. Making your own spawn requires extreme attention to detail, good knowledge of the needs of that specific species, and the facilities to work under sterile conditions. You're more likely to have success with spawn propagation once you have had some experience simply growing mushrooms. When producing mushroom spawn, unwanted germs (bacteria, mold, etc.) must be kept out at all times.

OUR TIP: *For those who would like to immerse themselves in the world of mushrooms more deeply or have experience in microbiology, the chapter entitled "Propagating Mushrooms" (p. 96) provides instructions for propagating mushroom spawn as well as an overview of the required materials and methods.*

How does mushroom spawn actually come to be? The first step in producing mushroom spawn is the acquisition of a pure mushroom strain. This can come either from spores or a tissue sample that has been transferred to nutrient agar, where a dense mat of mycelium forms, and then is transferred again to sterilized corn. Each species has its own particular preferences and needs regarding specific nutrients, so natural additives like gypsum or bran may have positive effects on the vigor of a given mushroom spawn. In general, nutrient agar provides for the nutritional needs of most types of mushrooms for initial mycelial growth. If things have been kept sterile, you will quickly observe your fungus culture thriving in its growing medium.

Mushroom spawn on a growing medium of grain

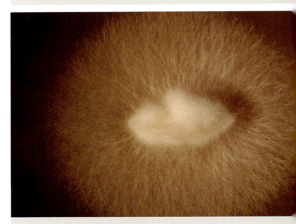
Fungal hyphae growing on agar in a petri dish

Mycelial fuzz growing on hardwood dowels

If the culture is somehow compromised, this should also quickly become apparent. Commercial spawn should also be closely observed for signs of health and vigor or contamination and disease. In general, mold fungi (greenish-blue spots) should never appear in healthy, thriving spawn. Modern commercial spawn is fairly reliably free of mold contamination. One thing is certain, however: Mushroom spawn that has been stored too long and/or at too high temperatures will have had its quality compromised. Storage for two to three weeks in a refrigerator or a cool basement or cellar is typically unproblematic.

Once the sterilized corn has been grown through with mycelium, it can be used to inoculate the growing medium that is intended to produce mushrooms. As previously stated, wood, straw, and compost are all appropriate for this purpose. In addition to sterile corn, hardwood furniture dowels—known in the mushroom world as "plugs"—can also be used as a growing medium for later inoculating logs. Sterilized, moistened dowels are inoculated with grain spawn. The mycelia colonize these plugs, which are then hammered into holes that have been drilled in logs.

OUR TIP: *Plugs are excellent for inoculating tree stumps. For more, see the chapter entitled "Cultivating Mushrooms in Woodlands and Fields" (p. 67).*

How Cultivated Mushrooms Colonize Wood—Inoculation

There are many ways to inoculate logs with mushroom spawn. Three methods have come to be seen as the most effective and accessible for the home grower:

- Cutting kerfs in a log with a chainsaw and inoculating with grain spawn.
- Drilling holes in a log with an electric drill or angle grinder with an adapter and hammering in plug spawn.
- For thicker logs and/or tree stumps, drilling holes with an auger bit and using an inoculation tool with grain spawn to fill holes.

Grain or Plug Spawn?

The following table is a brief overview of the most important advantages and disadvantages of each type of mushroom spawn. The method of inoculation (kerf, auger, plug) follows from the type of spawn used (grain, plug).

	Pros	Cons
Grain spawn (kerf method, auger method)	• large amounts of mycelium and large nutrient reserve • mycelium has large surface area or many boreholes through which to colonize wood • excellent cost-benefit ratio • time savings when inoculating many logs • readily available commercially	• limited shelf life (keep refrigerated) • susceptible to pests and mold • requires chainsaw
Plug spawn (plug inoculation)	• no chainsaw necessary; an electric drill will suffice • highly suitable for small-scale production or simply testing the waters of mushroom cultivation • Pest problems greatly reduced (no mice, no germs, no fungus gnats) • spawn stores well	• comparatively smaller mycelial mass, requiring many more boreholes • more time required for spawn to colonize log • later development of fruit bodies • more labor intensive

Mushrooms from Logs

Mushroom cultivation using logs is fairly simple and high yielding. There are many wood-loving cultivated mushrooms that you can easily grow in your own garden. One of the world's best-loved edible mushrooms, shiitake, is also one of the best to grow on logs. This highly natural cultivation method has long been practiced in Asia and is now also practiced in other places such as Europe and America. Other mushrooms that are cultivated on logs include sheathed woodtuft, various oyster mushrooms, enokitake, and also sulphur tuft. Cultivating on logs approximates these mushrooms' natural life cycles. Most of today's commercially available mushrooms are produced by highly technical means under strictly controlled conditions in vast halls. In order to provide for optimal growing conditions, you will need to keep several factors in mind.

Wood Quality

To successfully grow mushrooms, you need high-quality wood. Wood for mushroom growing should be fresh, not more than four months old. Ideally you will be able to use wood from trees felled in winter or very early spring; because deciduous trees are dormant in winter, their sap is not flowing and the sugar content of their cells is at its highest. This high sugar content of the sapwood encourages the growth of mycelium in the trunk and is important for rapid colonization. Plus, wood that has been stored longer is more likely to have been "infected" by competing, undesired fungi, which can make it difficult if not impossible to establish your desired cultivated mushroom species.

It is possible to cultivate mushrooms on both hardwoods and softwoods. However, hardwoods like common beech (*Fagus sylvatica*), hornbeam (*Carpinus betulus*), oak (*Quercus* spp.), and maple (*Acer* spp.) can yield mushroom harvests for up to five years, whereas softer woods like willow, birch, poplar, alder, and linden only produce mushrooms for at most three years. At the same time, since mycelia spread through softwoods faster than hardwoods, they will produce mushrooms sooner.

Beech, oak, and birch are the most common woods used for mushroom cultivation in central Europe. Beech and oak logs are used in shiitake production, as the properties of these woods match up well with the needs of shiitake mycelium. Sheathed woodtuft and lion's mane, on the other hand, do well on birch. The ideal wood for each mushroom is described in the Mushroom Portraits (p. 42 and 64).

OUR TIP: *Our recommendations for which species of wood to use for each mushroom are based on our many combined years of experience and trials. We are not aware of any noteworthy positive examples of cultivating edible mushrooms on fruit tree wood.*

Bark Quality

When felling trees, it is important that the bark is not damaged. When the whole bark remains on the wood with no nicks or other injuries, the wood stays moist and potential fungal competitors have the least amount of surface area exposed to penetrate the wood. Furthermore, whole, uninjured bark is less likely to fall off early.

Why is healthy bark so important for successful mushroom cultivation? The bark protects mycelium from drying out. A log can appear pretty well spent, but as long as the bark is still intact, mushrooms can still sprout. Beechwood is especially impressive in this respect. An old oak log after years of mushroom yields is another story, however. From the outside one hardly notices any signs of age: the bark does not fall off; it appears unaltered. Once you pick it up, however, you immediately notice how light it is, like wood you might find in a rotten, fallen tree trunk. Whatever wood is used, sooner or

CULTIVATING MUSHROOMS IN THE GARDEN

Beechwood logs

later the usable portions of the log are used up by the mycelium and it must be replaced by inoculating new, fresh wood.

OUR TIP: *For active mushroom gardens, we recommend inoculating new logs every two to three years to guarantee a harvest every year.*

Sources of Wood

Of course, not everyone has the luxury of owning their own forest from which you might harvest wood for mushroom cultivation as you please. In procuring wood for inoculating with mushroom spawn, it is perhaps easiest to simply ask a farmer in your area if he or she would be willing to set aside a few feet of hardwood for you when preparing firewood in winter. Be sure to mention the diameter you'd like the logs to be and how important it is that the bark remain intact. 4–6 in (10–15 cm) is a typical diameter for shiitake cultivation. At the other extreme, sheathed woodtuft can colonize logs up to 14 in (35 cm) in diameter.

Before Inoculating

Store logs covered outside until spring. Wood felled in winter and early spring can be inoculated in mid to late spring.

Important: When using oak logs to grow mushrooms, make sure they sit for three to four weeks before inoculating. This has the effect of reducing the concentration of naturally occurring antigens in the wood. This is how a healthy tree protects itself from potentially harmful microorganisms. If you do not observe this waiting time, the mycelium will have a much more difficult time in colonizing the log. Also, make sure your logs are sufficiently moist when inoculating. Cracks on the cut ends of logs are indicators of excessive drying. If your logs are too dry, submerge them in water in a clean container for 24 hours (longer risks adversely affecting mycelial growth by soaking up too much water, thus purging oxygen). Thinner logs (3–6 in [8–15 cm]) are especially prone to excessive drying and are thus more likely to require soaking before inoculation.

Covered wood typically holds its moisture for several months and does not require any soaking or watering. A simple woven polyester tarpaulin can be used for covering. Do not store wood in the sun, as even springtime sun can be intense enough to be excessively drying.

Inoculation with Mushroom Spawn

Kerf Method

This is a method for inoculating about 3-feet-long (1 m) logs with mushroom spawn. Cut a kerf (a slit or channel made by cutting with a saw) about 1 foot (30 cm) from each end, so two kerfs per log, one from above, one from below. Shiitakes can be cultivated with slightly closer kerfs of 9–10 in [23–25 cm] which means three kerfs will fit on each log.

The kerfs should be just wide enough (5/8 in [1.5 cm]) to allow for filling with spawn. Tape (ideally duct tape) is used to seal the spawn into the kerf, so you may find it helpful to tape up the kerf site up to two widths of tape wide

20 HOME-GROWN MUSHROOMS FROM SCRATCH

Kerf method of log inoculation

and completely around the log before sawing, so the final piece of tape has something to adhere to.

If you haven't started with taping the log, tape up the kerf and then cut out a small rectangle. Backfill with spawn as full as possible, using a clean flat stopper of some sort to pack it into corners and fill empty spaces, then seal with tape. See to it that the spawn is not too tightly squashed in; it should just be packed enough to eliminate air pockets. When grain spawn (see Making Grain Spawn, p. 103) has been compactly and cleanly packed in, fungus gnats and other pests and diseases are much less likely to thrive. For rougher-barked woods like oak, take extra care with the tape to maximize air- and water-tightness of the kerf, which minimizes access to the spawn for pests and contaminants as the fungus colonizes the log and holds in moisture. To prevent the tape from coming off, use staples to further secure it to the log.

Cut kerfs and inoculate all in one session to prevent wild fungi and other organisms from contaminating the log and beating your spawn to the finish in colonizing the logs. If you are using multiple edible mushroom species, never inoculate the same log with more than one kind of spawn, as they will compete with each other and only one will ultimately colonize the log.

There is no reason to worry about your cultivated mushrooms spreading to and infecting fruit trees (or any other trees for that matter). Normally, the natural defences of a healthy tree are so strong that even the increased presence of spores in the air does not lead to infection. How many logs can be inoculated with 2 quarts (2 liters) of grain spawn? It depends on the diameter of the logs and the type of mushroom. For shiitakes, up to 6 to 8 stacked logs or 2 to 3 buried logs (3 ft/1 m long) can be inoculated.

Labeling

Remember to Label

- Label logs immediately after inoculating. Aluminum tags have shown themselves to be the most durable and reliable material for labels
- Note species of mushroom and year of inoculation on the label
- Affix label to end face of log with staples
- Do not use highlighters or other markers; these simply do not last

Required materials for the kerf method:

- grain spawn
- chainsaw
- wooden stick (such as a paint mixer) for packing spawn
- duct tape
- utility knife
- staple gun with staples

Overview of individual steps in the process:

- procure logs (in the winter months)
- procure grain spawn (mid to late spring)
- assemble required materials

Inoculation procedure (see Kerf method, p. 19):
- cut kerfs into 3-foot-long (1 m) logs with the chainsaw (three kerfs for shiitake, two for all other species)
- cover kerfs with duct tape (or, if you prefer, cover kerf site before cutting)
- cut out rectangle in tape with utility knife
- pack grain spawn into kerf until full, tightly enough to eliminate air pockets, lightly enough to not squash spawn
- close fill hole with tape
- anchor tape onto wood with staples
- stack in a shady, wind-protected spot (see photo of logs stacked for winter, p. 25)

Plug Spawn Method

Plug spawn is extremely practical for inoculating smaller quantities of logs. The method involves drilling regularly spaced holes in logs using a drill or an angle grinder with a special adapter. You then place plugs in the holes and hammer them home. Depending upon log diameter and mushroom type, you will use 25 to 50 plugs for shiitake or 50 to 100 plugs for other types per 3 feet (1 m) of log length. In general, the more plugs per log, the more successful the colonization. It is critical to seal each inserted plug with a special plug wax.

How to actually proceed? Use a drill with a $1^{1}/_{32}$ or $^{3}/_{8}$ in (9 or 9.5 mm) bit (adapters for angle grinders that are specifically designed for boring holes for plug spawn are also available) to bore regularly spaced holes 2 in (5 cm) deep into each log. Do not use a bit smaller than $1^{1}/_{32}$ in (9 mm) as the smaller holes they create may cause the plugs to be stripped of mycelium when hammering in. Good plug spawn consists of well-colonized, moist hardwood dowels, which are highly effective in colonizing fresh logs. Place plugs in the holes you've

Inoculating a log with plug spawn

bored, hammer them in, and seal with melted wax applied with a brush.

If no plug wax is available, so-called cheese wax can be used. We do not recommend using beeswax, as this becomes porous at low temperatures and will fall off. The purpose of the wax is to hold in moisture. To apply, first melt the wax, which can easily be done in an appropriate container with a candle or tea light. Apply generously to each inserted plug with a brush.

It is worth noting that fresh plug spawn can be stored for up to two months with no loss in quality. As long as it is not opened, it cannot become contaminated and its moisture content remains constant. Sometimes small mushrooms even start to appear on the plugs, which should be removed before using to inoculate.

> How many logs can you inoculate with 100 plugs? It depends on the diameter of the logs and the kind of mushroom. For shiitake: a 3-ft (1-m) log, 4–6 in (10–15 cm) diameter, will take 50 plugs. For all other mushrooms: a 3-ft (1-m) log, 8–10 in (20–25 cm) diameter, will take a minimum of 80 plugs; a 3-ft (1-m) log, 10–12 in (25–30 cm) diameter, will take a minimum of 100 to 120 plugs.

Required materials for the plug spawn method:

- plug spawn
- drill with $1^{1}/_{32}$ or $^{3}/_{8}$ in (9 or 9.5 mm) bit or angle grinder with plug-hole-borer adapter
- plug wax for sealing (often included with your plug order)
- tea light or cooking plate for melting wax
- hammer

Materials needed for inoculating with plug spawn

Overview of individual steps in the process:

- procure logs (in the winter months)
- procure plug spawn (mid to late spring)
- assemble required materials

Inoculation procedure (see Plug Spawn Method, p. 22):

- drill regularly spaced holes
- hammer plugs into holes
- melt plug wax
- brush wax onto each inserted plug
- stack in a shady, sheltered spot see photo of logs stacked for winter, p. 25)

OUR TIP: *Kids are great help when inoculating with plug spawn—they're usually enthusiastic, interested, and have tons of fun.*

Auger Method

The auger method is excellent for inoculating tree stumps or logs.

An auger is used to bore holes 3–5 in (8–12 cm) deep at 2–3 in (5–7 cm) intervals into a tree stump. You can bore into either the cut face or the side of the stump. We recommend a $5/8$ to $7/8$ in (16–22 mm) auger. Be sure to make plenty of holes—10 to 20 holes will be needed for a mid-sized stump. Crumbly grain spawn is then shoved in each hole with a stick and then the holes are closed with wood or cork stoppers.

OUR TIP: *Hardware stores stock long round wooden rods. These can be cut into pieces and used as stoppers. Additionally, use wax to seal holes. A funnel makes it easier to get spawn in the holes.*

It is important to not wait too long to inoculate—the older the stump, the lower the likelihood of success. Ideally, the stump should be from a tree felled three months earlier or less. After about three months, other wood-decomposing fungi have likely colonized the stump, making colonization with an edible mushroom difficult if not impossible. Moss on the cut face of the stump indicates that the window of opportunity has closed.

In the Mushroom Portraits (p. 42 and 64) you will find a list of cultivated mushrooms that grow especially well in stumps.

Auger method of inoculating logs

Logs stacked for overwintering

The Development Stages of the Mushroom Garden

Colonization and Storage

After inoculation, mycelium gradually grows through the logs. Ideally, this takes place from spring through autumn. When inoculating in the spring, winter-felled wood is perfectly primed for success in that it should still be moist and has not yet been colonized by wild fungi.

The leftover stump can then be inoculated along with the logs cut from the rest of the tree. The time required for colonization varies by mushroom type, wood type, diameter of the logs, and the number of dowels or kerfs.

Oyster mushrooms take over logs fairly quickly compared to other edible mushroom varieties. In general, though, plan on colonization taking a year. Shiitake can need up to two years, though with the kerf method it will be closer to one.

Once you have inoculated your logs, they should be stacked in a shady location in the garden. Do not stack logs directly on the ground; keep them at least 4 in (10 cm) off the ground. Cover your stack of inoculated logs with a tarpaulin. For larger stacks, take the additional precaution of protecting them from the wind.

OUR TIP: *Some books on mushroom cultivation recommend packing straw into the log stack. We have had bad experiences with this. The straw does help speed the colonization process along, but the bark is negatively affected. It falls off more easily, which reduces yields and shortens the life of the logs.*

When dry weather predominates over a long period of time, the log stack will need watering. Also, check your logs from time to time to ensure that mice have not discovered the spawn contained in them. If necessary, protect inoculation sites on the logs with finely woven hardware cloth or aluminum tabs. Logs stored outdoors in winter should be covered.

Establishing a Mushroom Garden

When selecting a location for your mushroom garden, consider the following: Mushrooms do best in a moist microclimate. The site must be mostly shady—ideal locations include under fruit trees, bushes, and other locations that tend to stay moist (ferns and moss are good indicators of this). Once you've selected a site, you can start designing and arranging your mushroom garden. Make sure you allow for sufficient space to store your inoculated logs.

The first step in "winterizing" logs inoculated by the kerf method is to remove the tape from the inoculation sites. Now you will be able to judge how successful the inoculation has been. Often you will be able to see the white mycelium of your mushroom culture on the ends. The grown-in spawn will look mostly white, though white/brown spots are normal for shiitake.

Sometimes grain spawn can get contaminated by other organisms. However, part of the spawn becoming a bit darker is no cause for concern—white, healthy spawn can usually be found underneath, and your chances for successful colonization remain high.

Shiitake logs require no contact with the ground and are to simply be left as they are (3 feet [1 meter] long). For most other species, proceed as follows: Cut each meter-long log into three equal pieces. Then, bury these 4 in (10 cm) deep at the location you've selected for your mushroom garden. Oyster mushrooms, sheathed woodtuft, and others need contact with the soil in order to form fruiting bodies. To bury your logs, simply dig each hole with a diameter that corresponds to the diameter of the log to be placed in it, insert your log, then backfill as necessary and compact the soil. After about three weeks you will again be able to check for successful colonization by tipping a log slightly. If you see white fungal growth, that means your mushroom's mycelium is growing into the ground (see photo, p. 27, left). It is a characteristic

Cut right next to kerf when dividing logs to be buried

Well-established shiitake mycelium

Log cross-section: kerf method

Log cross-section: plug spawn method

Log cross-section: auger method

CULTIVATING MUSHROOMS IN THE GARDEN

Mycelium spreading into the soil

Stapling moss onto the tops of buried logs

of these species that mushrooms also grow out of the ground surrounding the log and take up water and nutrients from these areas as well. For logs inoculated with the kerf method, make sure that the two end pieces of the log you've divided in three have been buried with the inoculation site on the bottom, buried in the ground. This helps mycelium grow faster into the ground and minimizes the inoculation site's exposure to pests.

OUR TIP: *Covering the exposed cut face of buried logs with moss helps them retain moisture, improves conditions for your edible mushrooms, and looks good, too. Collect moss from rocks in the forest. Thin moss with a minimum of humus and detritus is preferable to tall, bushy moss growth. Attach moss to logs with staples and water it several times soon thereafter.*

Eventually it will grow onto the log and act as a living roof, providing shade and retaining moisture. *The moss is also a good indicator of the water requirements of the log—logs sometimes need to be watered during prolonged dry periods in high summer. If the moss atop the logs appears fresh and moist despite drought, there is no need to water the logs. This is especially relevant for summer fruiting mushrooms.*

Moss can also be used for logs inoculated with plug spawn.

Keep the area selected for your mushroom garden free of tall grass to help keep slugs and other pests away, and regularly remove the grass immediately surrounding your buried logs. Logs can be buried with spacing as little as 6 in (15 cm) from one another, a spacing that minimizes the overall area of the mushroom garden while still allowing sufficient space to harvest comfortably. Of course, you can also choose to bury logs here and there throughout

Logs stored horizontally

Logs can be hung from hooks

Herbert Wurth hanging shiitake logs

the garden as long as shade, moisture, and wind protection are sufficient. For shiitake logs, there are two typical approaches to situating them:
- Horizontal: make a stand by pounding four stakes or posts into the ground and attaching two boards or logs atop them. Lay the logs across the boards.
- Vertical: drill holes in the logs and hang them from a tree branch or stand.

Each approach protects the logs from slugs. If there are absolutely no slugs in your garden, you can simply lean shiitake logs against a tree trunk. Mushrooms that require soil contact are highly susceptible to damage from pests (like slugs). In the chapter entitled "Pests and Competing Organisms in Mushroom Cultivation" (p. 90), you'll find lots of advice on how to protect your mushroom cultures.

Watering Mushroom Logs

It helps to water logs often after burying them to "jump start" your mushroom culture. The moss and soil contact will eventually take over the moisture regulation of the logs. Excessive watering is problematic, however. Watering cans and/or a garden hose will be useful here. For larger mushroom gardens, more involved irrigation systems may make sense, such as a sprinkler or a misting system. There are two main approaches to consider in the event of a heat wave in which you are unable to water your mushrooms:

CULTIVATING MUSHROOMS IN THE GARDEN

Mist sprayer irrigation system

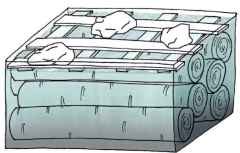
Submersion tub for shiitake logs

- Since most mushrooms adjust well to the rhythms of nature, they may be just fine without watering. Mushrooms that normally thrive in summer may just have a lower yield. When mushrooms appear on buried logs, they should be watered every day until just before full ripeness. In wet years, no watering will be necessary.
- A second possibility would be to use an irrigation timer (especially when growing shiitakes, which are especially sensitive about water). This may be more labor-intensive and more expensive, but if you incorporate a vegetable or flower garden into the system it may make it worth it.

Shiitake's Uniqueness

Shiitake is handled differently to those mushrooms cultivated with soil contact. This mushroom's native climate is very humid and it requires lots of water while fruit bodies develop and during harvest. In fact, shiitake requires a unique procedure to stimulate fruiting. In mid and late summer, submerge logs completely in cold water for 24 hours. Typically, one or more tubs that are large enough to hold all the logs to be submerged are used to this end, along with some sort of weight (such as a stone or boards) to hold down the logs that would otherwise float (thus exposing some of the logs to air, which would render the technique ineffective). Logs could also be submerged in a pond, brook, or other natural body of water, as well as in a rain- or water-filled wine barrel. The logs then absorb water until they are saturated—an important first step for a successful harvest. After submersion, knock the logs hard on the ground three or four times, as the physical shock helps "wake up" the culture and induce fruiting.

Following the submersion procedure, water your logs daily to keep them saturated. This is especially important when tiny fruit bodies begin to emerge through the bark. If the log dries out at this crucial moment, the mushrooms cease growing. Otherwise, in high summer, the logs can simply remain in their normal storage positions without any special attention. Between harvests, shiitake logs need time to rest, during which they should dry out. The steps described above should be repeated every year. Once mushrooms have reached full size, watering should be reduced so the mushrooms do not become spongy and bloated.

HOME-GROWN MUSHROOMS FROM SCRATCH

"Waking up" the shiitake culture

A "mushroom garden" of buried logs

Crucial points to keep in mind:

- use cold water
- submerge for a full 24 hours, but no longer
- do not forget to knock logs on the ground afterward
- water regularly (while mushrooms grow)

Harvesting Mushrooms

The incubation time and harvest season varies from one kind of mushroom to another. You will need to be patient, as it can take up to two years for the first mushrooms to emerge from your logs. There are biological processes at work that simply need time to complete. Golden oyster flushes with mushrooms several times in the summer months, sheathed woodtuft in autumn (sometimes also in spring), and blue oyster in autumn. Shiitake has two harvests per year, one in late spring/early summer and one in late summer/early autumn. In wet summers, mushrooms may appear on their own, without any additional watering necessary.

Shiitake needs to be submerged and knocked, however, even in wet years, as waterlogged conditions and intense vibration are crucial to induce fruiting. Depending on the diameter of the logs and type of mushroom, the logs will produce mushrooms for several years (up to five) and will produce as long as their nutrient reserves last. Hardwoods contain more stored nutrients and thus last longer (four to five years). For more information and for information about post-harvest handling of mushrooms, see the chapter entitled "Recipes and Processing Edible Mushrooms" (p. 118).

The Stages of Mushroom Cultivation Throughout the Year

To help you keep track of the various tasks of mushroom cultivation and when they are to be carried out, we have summarized them in this table:

Year	Winter	Spring	Summer	Autumn
First	Wood procurement: winter- or early-spring-felled trees	• Inoculation: procure grain and/or plug spawn • Inoculate 3-foot-long (1 m) logs • Store inoculated wood covered in stacks until following spring	• Colonization phase: fungal mycelium grows through wood • Inspect for chewing damage from rodents	
Second	Period of rest Harvest: Enokitake often come in the first winter after inoculation	Mid spring: Cultures with soil contact: • remove tape • divide meter logs in three • bury logs 4 in (10 cm) deep, inoculation site face down • cover tops with moss Mid to late spring: • unstack shiitake • remove tape Late spring: • submerge shiitake 24 hours • knock • lay out or hang • water and harvest	Harvest: golden oyster mushroom, Italian oyster mushroom	Harvest: sheathed woodtuft, blue oyster mushroom, shiitake (again 24 hours of submersion!) After harvesting: leave buried cultures in the ground; stack shiitake
		Harvest: shiitake until early autumn		
Third	Same cycle as for second year (logs outside entire year, etc.)			

Harvest Months for Mushrooms Grown on Wood

Mushroom/Month	1	2	3	4	5	6	7	8	9	10	11	12
Shiitake						x		x	x			
Golden oyster						x	x	x				
Blue oyster				x	x					x	x	
Italian oyster						x	x	x				
Late oyster				x	x					x	x	
Sheathed woodtuft				x	x					x	x	
Nameko										x	x	
Enokitake	x										x	x
King stropharia						x				x		
Lion's mane							x	x				

32 HOME-GROWN MUSHROOMS FROM SCRATCH

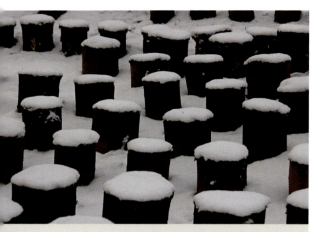

Snow-covered mushroom logs

Herbaceous perennials in the mushroom garden

Cultivated Mushrooms as Features of Garden Design

Mushrooms cultivated on logs are now starting to make their way into ornamental gardens. Gardeners have begun to see how buried oyster mushroom logs can provide accents in beds, borders, and raised beds. Why not use mushroom logs to make your own garden more ornamental, too? Laid out in a circle or in a spiral, moss-covered mushroom logs make the garden more interesting all year round, even in winter. And integrating cultivated mushrooms into the garden is not just worthwhile for culinary reasons; they are also a great way to use shady, wet parts of the garden to your advantage. In this way, "problem corners" of the

Combining Mushrooms with Shade-Tolerant Plants

Moist Shade	Arid Shade
plantain lily (*Hosta*) wood fern (*Dryopteris*)* false spirea (*Astilbe*) lady's mantle (*Alchemilla*) Christmas rose (*Helleborus*) bulbs: snowdrop (*Galanthus*), spring snowflake (*Leucojum*), fritillary (*Fritillaria*) geranium (*Geranium*)* bell flower (*Campanula*)* goat's beard (*Aruncus*) great wood rush (*Luzula*) peppermint (*Mentha*) bleeding heart (*Lamprocapnos*) hydrangea (*Hydrangea*) ivy (*Hedera*)* yellow archangel (*Lamium*) columbine (*Aquilegia*)* yellow flag (*Iris*)	periwinkle (*Vinca minor*/*Vinca major*) Alpine barrenwort (*Epimedium*) liverwort (*Hepatica*) lungwort (*Pulmonaria*) elephant's ears (*Bergenia*) tulip (*Tulipa*) Note: On dry sites, it may be necessary to water your mushroom logs. The more lavish the growth of companion plants, the easier it is to maintain a moist microclimate for your mushrooms.

* also appropriate for arid shade

CULTIVATING MUSHROOMS IN THE GARDEN   33

garden can be transformed and teem with life. Combined with shade-tolerant plants (see table opposite), mushrooms flourish in these kind of sites and they also contribute to diversity.

The plants listed in the table are excellent for greening up your mushroom cultivation site. There are any number of combinations possible with these species.

Companion Plants for the Mushroom Garden

Shade plants offer interesting leaf shapes and structures, which when combined with flowering plants make for a harmonious effect. Waterlogged soils are good for neither mushrooms nor any of the shade-tolerant plants listed here. Ferns and grasses are excellent for providing the basic framework of a shade planting. Spring-blooming tulips combined with bulbs contribute to a changing display while the tender green of yellow archangel and periwinkle can be used to cover the ground. In this kind of low-growing vegetation, mushroom logs can create attractive accent notes.

Mushroom Portrait: Luminescent panellus *(Panellus stipticus)*

Inoculated logs can fulfill a second function: defining garden paths. A mushroom that is particularly interesting to this end is luminescent panellus (*Panellus stipticus*). This mushroom, as its common name suggests, has developed the incredible ability to glow in the dark. In our native central Europe, roots infested with the honey fungus (*Armillaria mellea*) sometimes glow in the dark in a phenomenon known as bioluminescence, which is a well-known feature of glowworms. In early history, wood illuminating with the glow of luminescent panellus was a mysterious, eerie phenomenon that was associated with magicians, fairies, and elves. In addition to the fruit body, the mycelium of *Panellus* also

Luminescent panellus by day

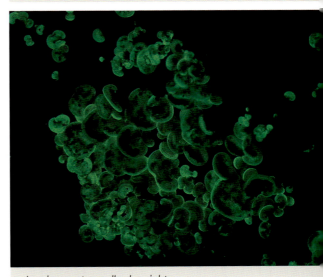

Luminescent panellus by night

Luminescent panellus (*Panellus stipticus*)

Recommended growing medum: wood	birch, beech, oak
Log diameter	Luminescent panellus is typically grown on logs with a diameter of 8–10 in (20–25 cm), 3 ft (1 m) in length, with plugs
Average colonization time	• 1 to 2 years • after colonization: cut log in thirds, bury, cover with moss
Fruit body development	summer through autumn
Specific features	• logs to be buried • mycelium and fruit bodies glow • other uses: for wood in terraria
Decorative feature	Fruit bodies form on inoculated hardwood dowels stored in a sealed jar (do not open). In this isolated environment, bioluminescence can be observed for several months. We recommend keeping such a jar on your nightstand. Mushrooms grown in this way appear different from those grown on logs thanks to the higher carbon dioxide levels in the jar.

Fence construction for shiitake logs

glows. Interestingly, only eastern North American strains of *Panellus stipticus* glow. European *Panellus stipticus* is called bitter oyster, astringent panus, or stiptic fungus and does not glow. The mushroom can be grown decoratively in the garden but is not edible. It is creamy-white in color and has ½ to 1 in (1–2 cm) muscle-like fruit bodies that grow along the entire stem. New fruit bodies appear every year.

Shiitake Logs as Privacy Fence

Shiitake logs can work well as a privacy fence or in partitioning different parts of the garden. Three-foot-long [1 m] logs can be hung from a wooden framework, suspended as high or as low above the ground as you prefer. We recommend spacing of at least ~3 in (7 cm) between hanging logs to allow space for mushrooms to grow.

OUR TIP: *With an appropriate drill, bore a hole in each log and use this to hang logs from your construction with a cable (e.g., clothesline with wire core). A simple and effective method for attaching to a beam is to drill a hole in the beam every 10 in (25 cm). Thread the clothesline through the*

CULTIVATING MUSHROOMS IN THE GARDEN

A decorative design element in your garden . . .

. . . allows special glimpses beyond

hole and use ⅝ in (1½ cm) diameter, 1½ in (4 cm) long wooden rods to secure (see photo opposite). To submerge the logs, simply remove the rod and later use it again to hang the logs back up. If uniformity is important to you, use logs of equal diameter and length.

Designing a unique, flourishing, and well-nurtured ornamental mushroom garden will bring pleasure year after year.

Mushrooms from Straw

The inoculation of straw bales with mushroom spawn is the perfect complement to mushroom cultivation on logs. Because fungi are able to colonize straw much faster than wood, you can expect a mushroom harvest in the same year. While mycelia are still growing through logs, you can be harvesting mushrooms from straw-grown fungi. Bales can be harvested for about four months, after which all of the nutrients required by the mushroom culture will have been used and new bales should be inoculated.

Requirements for Cultivating Mushrooms on Straw

We recommend simply cultivating on bales situated in the garden. Select a site that is easily accessible, is easy to water, and has a shady, moist microclimate. In general, the same siting rules that apply to logs also apply to straw bales. Lay straw bales directly on the ground so the mycelium has access to the soil. If your site is too windy, the straw bale may dry out, and if your site is too cold, the mycelium will colonize the straw bale more slowly. Good growing conditions are often found under deciduous trees or along hedges. Cold frames and greenhouses can also be appropriate sites, as long as sufficient shade can be provided. Keep temperatures within the straw bale under 86° F (30° C), as mycelium will begin to die off above this limit.

A large array of delicious edible mushrooms can be cultivated on straw, including king stropharia (*Stropharia rugoso-annulata*), also known as wine cap stropharia, garden giant, and burgundy mushroom. Other popular types include various oyster mushrooms

Cultivate on Wood or Straw?

The following table lists the most important advantages and disadvantages of each growing medium:

	Advantages	Disadvantages
Wood	Large nutrient reserves Many years of production Usually competition-free, as freshly-felled wood is sterile Simply overwinter in garden	Long colonization phase (~1 to 2 years)
Straw	Quickly colonized Easy to inoculate Spawn easy to incorporate Harvest within months of inoculation Straw can be composted after nutrients for mushrooms are spent	Dries out relatively easily Competition from organisms already established on the straw Shelter and reproduction site for slugs Wetting and fermentation of straw bales is tedious

Freshly inoculated straw bales

Small, young king stropharia

such as pink oyster, golden oyster, blue oyster, and Italian oyster. Each mushroom type has its own favorite temperature that you should be aware of. Mycelium grows in a temperature range of 41–86° F (5–30° C), summer-fruiting mushrooms prefer 77° F (25° C). For those who decide on blue oysters (an autumn and spring mushroom), a mere 41–59° F (5–15° C) will suffice for fruit body development. Inoculate either in spring (harvest later that year) or in autumn (harvest the following year).

Straw Quality

The quality of the straw is an important factor for successful colonization. The straw should never have been handled with fungicides, as this may have a negative influence on the yield and quality of mushrooms. Another important prerequisite is a source of organically grown straw bales. There are a few indicators of quality straw bales that you should observe. Healthy straw is golden yellow to golden brown in color. The straw should be strong and difficult to rip apart. Make sure there is a minimum of material from weed or companion plants in the bales, as this contributes to mold growth. Moldy or old straw should not be used and often has an unpleasant odor and has visible grayish-black spots. Bales weigh around 22–33 lb (10–15 kg). Wheat, rye, and barley straw can be inoculated with mushroom spawn. Oat straw's structure makes it a less ideal candidate for inoculation. The more pressure that was used to bale the straw, the higher the yield will likely be. When shopping for bales, make sure they have been stored in dry conditions.

OUR TIP: *Shop for straw bales in late summer, after that year's grain harvest. Availability should be higher and the straw will be fresher.*

Preparing Straw for King Stropharia

Before inoculation you will need to water your bales extensively. Without this step, mycelia are unable to spread through the straw. There are multiple possible methods to accomplish this:

- Submerge an entire bale underwater in a sufficiently large tub, barrel, or other container. Use a weight to keep the entire bale underwater. Keep the bale entirely submerged for 48 hours. Important: Locate the submersion barrel in the immediate vicinity of where the bales will ultimately reside, as fully soaked bales are many times heavier than dry bales before soaking. If possible, change the water once or twice.
- Another option is to water repeatedly with a can or garden hose. This is a more labor-intensive method than simply soaking. Bales should be watered several times a day for four to six consecutive days. Optimal wetness has been reached when the bales are wet all the way to their cores and water runs out from underneath. You can also test for wetness by removing a few pieces of straw from the middle of the bale. If several drops of water appear when squeezing together, the bale should be sufficiently wet.

OUR TIP: *Straw is naturally somewhat hydrophobic because of being surrounded by a thin layer of wax. This is why it takes so long to sufficiently soak the bales. Pouring hot water over the bales before soaking speeds the process. If there is no access to heated water, you will just have to go through a longer soaking process as described above.*

Inoculating with King Stropharia Spawn

After soaking, set out bales to drip dry for a day. This rids the bale of excessive water, making it ready for inoculation. Where possible wear work clothes for this, as wet straw can stain. To inoculate the bales, you will need to bore holes in them and fill them with spawn. A wooden stake, the end of a broom, or something similar can be used to create the hole. Holes should be 4–6 in (10–15 cm) apart, about 15 cm (6 in) deep and regularly spaced.

OUR TIP: *Grain spawn and straw spawn (mycelium on a growing medium of straw) are both available commercially. King stropharia straw spawn has shown itself to be reliable for straw bale inoculation. It has the advantage of being preadapted to the growing medium it is being*

Inoculation with grain spawn

used to inoculate. As a result this spawn colonizes faster and yields more reliably. For oyster mushrooms, grain spawn performs very well.

Two liters of spawn will inoculate about two straw bales. Before opening the grain spawn package, break up the spawn a bit by squeezing and rubbing the bag. This makes the actual filling of holes with spawn much easier. Wash your hands and begin filling the holes you've made in the straw bales with spawn (you may find a spoon to be useful here). Make sure the spawn gets packed all the way to the bottom of each hole; push all the way in with a rounded piece of wood if necessary. Do not fill holes completely to the top with spawn as you will ultimately close them with loose straw. Tap straw caps lightly with your foot to close well. Sealing well is especially important when inoculating in late spring, a time when mycelium needs extra protection from outside influences. In autumn and early spring, there are fewer pests and diseases that could negatively affect the mycelium and are thus the best times to inoculate.

Colonization Phase

As mycelium spreads its way through your straw bales, keep them in the garden. In strong or persistent rain, cover inoculated bales with a tarpaulin or plastic sheet, as too much water negatively affects the mycelium. Early to mid autumn are when autumn inoculations should take place. Inoculating later risks that the mycelium is unable to establish itself well enough before winter. In regions with harsh climates, covering bales with fallen leaves can help protect growing mycelium. Once the bale has been sufficiently colonized, frost no longer presents a danger.

If you're looking to inoculate in spring, pay special attention to the moisture level of the bales. As previously stated, too much water is not good for mycelial growth. It takes a bit of intuition and experience to get the conditions to be just right. If your site is getting too much sun, you can provide a roof for your cultures. In summer, water every four or five days with a watering can—for very shady locations, every seven to eight days should suffice.

Checking for Successful Colonization

Within three to four months of a spring inoculation, mycelium should have completely colonized the straw bales. Deep in the core of the bales, a white mycelial mat should be visible. The bales should have a pleasant mushroom smell. Small spots of mold are no cause for concern, as they will not have the foothold that king stropharia mycelium has on the bale.

Possible causes of cultivation failure:

- insufficient watering of bales before inoculation
- bales excessively populated with competing organisms to begin with (especially in the case of mold growth)
- pests (slugs, mites)
- waterlogging
- too little spawn used
- excessive watering (mycelium died)

Straw completely colonized by mycelium

If none of these apply to you, you can try covering the bales with potting soil (50 percent peat, 50 percent garden soil). First moisten the soil, then cover the bales with a 1½ in (4 cm) layer. Within four weeks, fruit bodies should have emerged.

Harvest

When inoculating in autumn, king stropharia should be harvestable in late spring or early summer the following year. Depending on weather, temperature, and extent of success of colonization, mushrooms could be harvestable earlier in an early, mild spring. When inoculating in early or mid spring, the mycelium will need two to four months to fully colonize the straw. The temperature needs to be above 50° F (10° C) for fruit bodies to form. The ideal temperature for fruit body formation is 68° F (20° C). One peculiarity about king stropharia is that it grows hidden in the straw and first appears as a fully ripe mushroom. There are several flushes of mushrooms, and harvest periods come in waves.

It is ideal to harvest mushrooms when their caps are still closed. Carefully twist them out, rather than cutting. Some mushrooms might ripen fully without emerging from the bale, so check for hidden mushrooms when harvesting by using your hands to spread the straw apart here and there so you can peek into the bale. And since the mycelium grows into the ground, mushrooms may grow out of the soil surrounding the bales.

Beware of slugs! These widely disliked creatures are attracted not only to your leafy greens and seedlings, but to your mushrooms as well. They hide in the bales and munch on mushrooms. Erecting a slug-proof fence at the start can save your mushrooms. Chemical slug protection is not advisable.

As soon as mushrooms become harvestable, keep the straw fairly wet. Water with a watering can whenever the outer 1 in (2 cm) becomes dry. See to it that there is nothing left of the mushrooms you harvest on the straw, as leftover mushroom pieces can attract fungus gnats and other pests.

After several flushes of mushrooms, the straw bale becomes exhausted of nutrients and mushroom flushes cease. A classic sign of this is the bale collapsing in on itself, somewhat like a raisin. When it's time to start a new culture, the entirety of the old bale must be removed from the growing site. You can either compost it or spread it in the garden as mulch.

Straw bale inoculation is the perfect counterpart to growing mushrooms on logs.

Cold frames can also be used to cultivate mushrooms

King stropharia covered with soil

The Stages of Cultivating Mushrooms on Straw Bales Throughout the Year

	Spring	Summer	Autumn	Winter
First year	Establishment: spring inoculation (early-mid spring) Acquire straw Procure spawn	Harvest: king stropharia, golden oyster, pink oyster As soon as first mushrooms form, keep bales wet Check for slugs	Establishment: autumn inoculation (early-mid autumn) Harvest: blue oyster Colonization: cover bales with fallen leaves against early frosts, cover bales with tarp in heavy rains	Straw bales remain outside, winter rest
Second year	Evaluate success of colonization	Late spring–early summer: harvest mushrooms inoculated in previous autumn As soon as first mushrooms form, keep bales wet Check for slugs	Harvest: late oyster, blue oyster Completely remove old straw bales—compost pile, garden mulch, etc.	

Cultivating Oyster Mushrooms on Straw

Preparing the Straw Substrate

King stropharia is not the only mushroom that thrives on straw; oyster mushrooms (*Pleurotus*) flourish on straw as well. The siting requirements are the same: shady, wind-protected locations are preferable.

Oyster mushroom culture has a unique setup. Straw bales are placed in a container with water and left for ten days without changing the water. Bales must be completely submerged in water, which you can accomplish by weighing down with a heavy rock. This causes the bales to ferment, which suppresses potentially competitive organisms and lowers the pH value. The relatively acidic conditions created through fermentation helps the oyster mushroom culture colonize the bales. The average outdoor temperature should be around 68° F (20° C) for optimal mycelial growth.

After fermenting, let water drain from the bales for at least 24 hours. As the bales ferment, they can begin to develop quite a strong smell, which is a sign of successful fermentation. After all this, the bales are finally to be inoculated (see Inoculating with King Stropharia Spawn, p. 37).

An overview of the steps of cultivation:
- submerge bales for 10 days
- drain bales for at least 24 hours
- procure grain spawn
- inoculate straw bales (bore holes, 15 per bale): fill holes with broken-up mushroom spawn (with clean hands and tools) and pack in by stomping with straw
- colonization phase

Colonization Phase

When establishing bales in spring, you will have to keep close tabs on the moisture levels of your bales in warm, dry weather. A rule of thumb for watering: When the outer inch (2 cm) of the bale is dry, water—up to 1 quart (1 L) water per bale should suffice.

The purpose of watering during the colonization and harvesting phases is simply to keep the surface moist. The core of the bales will still be sufficiently wet from being submerged earlier. During hot weather, bales can be watered

CULTIVATING MUSHROOMS IN THE GARDEN   41

Colonization Phase and Timing

The optimum timing for colonization of a mushroom culture depends on the species you intend to grow. In this table, possible timings for colonization and harvest are given.

Mushroom	Possible Colonization Time	Harvest
Golden oyster	Autumn, spring	Late spring–late summer
Pink oyster	Spring	Early summer–late summer
Blue oyster	Autumn, spring	Mid to late spring–early to mid autumn
Italian oyster	Spring	Early summer–late summer
Elm oyster	Spring	Late spring–late summer

more thoroughly, but with little water, several times per day. Changing weather can speed or slow the spreading of mycelium.

In order to improve the microclimate for pink, golden, and elm oyster, you can construct a small polytunnel over the bales. This can help the cultures colonize the bales faster and increase their yield. Make sure they get enough fresh air, however. The optimal temperature range for mycelial growth is 68–75° F (20–24° C). At temperatures above 86° F (30° C) in the middle of the bale, mycelium starts to die off. Check regularly for sufficient moisture in your bales. Oyster mushrooms are good to observe as they develop. Newly emerging mushrooms are highly susceptible to slug damage, which typically happens overnight. We recommend the use of slug-proof fences.

Harvest

Oyster mushrooms come in clusters. Harvest by cutting the entire cluster as each mushroom has one nutrient supply for the entire cluster. Removing one or a few mushrooms from a cluster causes what remains on the bale to die. The edge of the cap should still be somewhat tilted inward. Younger oysters taste best and are more "al dente" than older ones. Take care not to overcook and thus ruin their al dente consistency.

OUR TIP: *Mushrooms can be stored in the refrigerator for up to three days. A layer of mycelium may form on the surface of the caps, because the mushrooms continue to live after harvest. Such mushrooms are still edible. Cut at the base of the stem and remove growing medium if necessary.*

Oysters will produce several flushes of mushrooms with breaks of three to five weeks over the course of three to four months. Blue oyster fruits in a lower temperature range: 41–64°F (5–18° C). Frost does no damage—the mushrooms do indeed freeze, but recommence growing once it warms up again.

Once straw bales have shriveled up after the harvest is completely done, their remains can be added to the compost pile, used as mulch, or worked into the soil.

ting various growing media (straw bales,
edded straw, straw with wood chip layer)

Mushroom Portraits

Shiitake (Lentinula edodes)

This mushroom comes from Asia and was brought to Japan about five hundred years ago by Buddhist monks. Their cultivation was further refined in Japan and they ultimately made their way to America and Europe. In Traditional Chinese Medicine (TCM), it is an important medicinal mushroom. In any case, it is a wonderful edible mushroom that can be made into many different things.

Its cap is brown, the exact shade varying from strain to strain. Shiitake stems are tough but can be dried and pulverized for use as powdered mushroom. The white flesh of the mushroom is fairly firm. Shiitake has an unmistakable aromatic flavor and smell.

Oyster Mushrooms (Pleurotus)

Oyster mushrooms belong to a genus with a large diversity of individual species. What nearly all of them have in common, however, is that they are easy to cultivate. It is also interesting that they are not especially particular about growing media—they can be grown on logs, straw, or wood chips. When growing on logs, they require soil contact. When growing on straw bales or straw pellets, the straw must first be fermented.

Golden Oyster Mushrooms (Pleurotus cornucopiae)

This aromatic mushroom fruits in summer, usually multiple times. The golden oyster develops clumps of mushrooms on one stem and can grow to be quite large. Its golden-yellow color is similar to that of chanterelles and is a sight to behold. Slugs are also very much attracted to these, so be sure to take this into account when cultivating.

Shiitake

Shiitake	
Recommended growing medium: wood	Beech, hornbeam, oak, birch
Log diameter	Shiitake is typically cultivated on 3-foot-long (1 m) logs with a diameter of 3–6 in (8–15 cm)
Average colonization time	Inoculation with grain spawn: • kerf method or auger method: 1 year • plug spawn method: up to 2 years
Harvest	• early to mid summer; early autumn • produces mushrooms over the course of several years
Unique characteristics	• completely submerge logs for 24 hours before harvest • no soil contact
Culinary profile	Shiitake is a mushroom with firm flesh and a fine mushroom aroma and has innumerable uses in the kitchen (see Recipes and Processing Edible Mushrooms, p. 118)

Late oyster mushrooms

Golden oyster mushrooms

Blue oyster mushrooms

Pink oyster mushrooms

Blue Oyster Mushrooms (*Pleurotus columbinus*)

This mushroom is already fruiting by early to mid spring and then again in autumn. Frost does not damage it; mushrooms temporarily freeze, then resume growth when warmer temperatures return. The blue oyster mushroom's flesh is somewhat firmer than that of the Italian oyster mushroom, for example. It is a favorite in the kitchen thanks to its nut-like aroma. Furthermore, as with other oyster mushrooms, it can be used to help in biologically breaking down tree stumps (see Inoculating Tree Stumps, p. 75). It can vary from light gray to bluish dark green in color.

Late Oyster Mushrooms (*Pleurotus ostreatus*)

In the wild this tends to be found in deciduous forests, where it can be found on dead trees and stumps. Mushrooms quickly appear when temperatures reach the 39–59° F (4–15° C) range, which occurs late in the growing season, hence its name. Mushroom coloration can be a whitish light brown to grayish brown. This mushroom can enable you to have fresh mushrooms from the garden year round.

Italian Oyster Mushrooms (*Pleurotus pulmonarius*)

Yet another mushroom from this genus is the Italian mushroom, also called the lung mushroom (an anglicized version of its species name). It grows on wood from deciduous trees from spring to autumn. Its cap grows to $1^{1}/_{8}$ to $3^{3}/_{8}$ in (3–8 cm) in size and is often bent in along the edge, where depressions are found. Its color is white to grayish brown. The mushroom flesh is soft and has an aniseed aroma.

Golden Oyster Mushrooms, Blue Oyster Mushrooms, Late Oyster Mushrooms, Italian Oyster Mushrooms

Recommended growing medium: wood	Beech, hornbeam, birch, oak, maple
Log diameter	• oyster mushrooms are typically grown on logs with a diameter of 8–14 in (20–35 cm) • most find it easiest to inoculate 3-foot-long (1 m) logs and cutting these into thirds after colonization
Average colonization time	• 1 to 2 years • inoculation with grain spawn or plug spawn
Harvest	• golden oyster: several times in summer • blue oyster: autumn and spring • late oyster: late autumn, early spring • Italian oyster: several times in summer
Unique characteristics	• logs cut in thirds after colonization and buried 4 in (10 cm) into the ground • mycelium must grow into soil before fruit bodies can develop • mushrooms often grown from the area of contact between wood and soil • protect summer-fruiting varieties from slugs
Culinary profile	Oyster mushrooms are great for sautéing and in sauces. You may find it convenient to harvest when young and only use the caps.

Pink Oyster Mushrooms, Elm Oyster Mushrooms, King Oyster Mushrooms

	Pink oyster, elm oyster	King oyster
Recommended growing medium	Straw, straw pellets, mushroom kit	Mushroom kit
Environmental requirements	• straw must be fermented before inoculating • cultivation can proceed in garden (summer)	• high humidity • wet growing medium twice daily • grow in the house or in mini-greenhouse
Harvest	• spring to autumn, depending on time of establishment • several mushroom flushes spread throughout year	Constant harvest for up to three weeks after cutting open kit
Unique characteristics	Pink oyster: requires warm, moist climate	Produces a lot of fast-growing mushrooms
Culinary profile	• pink oyster: highly aromatic, turns orange-brown with cooking • elm oyster: fine flavor, excellent in sauces	Spicy flavor, firm flesh

Elm oyster mushrooms

King oyster mushrooms

Pink Oyster Mushrooms (*Pleurotus salmoneo-stramineus*)

This mushroom's color alone makes it a worthwhile addition to the garden. Its recommended growing medium is straw (see "Mushrooms from Straw," p. 35) as it only colonizes wood with difficulty. It is native to tropical and subtropical regions, and the prevailing conditions there (high humidity, warm temperatures) are required to grow it wherever you are. Its mycelium grows rapidly, which—as for the golden oyster—makes for high yields.

Elm Oyster Mushrooms (*Hypsizygus ulmarius*)

It can be a bit confusing that this is grouped with oyster mushrooms, as it does not belong to the genus *Pleurotus*. The elm oyster has become less common in central Europe; it can be found on deciduous trees such as elm and beech. Its growth habit as well as flavor is similar to those of other oyster mushrooms. It is white-gray to somewhat light brown in color and its flesh has a fine consistency. The best growing medium upon which to cultivate elm oyster mushroom is straw.

King Oyster Mushrooms (*Pleurotus eryngii*)

Flavor-wise, the king oyster mushroom is a delicacy. Fruit bodies can appear quite bizarre, as their caps sometimes hardly grow at all, making for some mushrooms that are all thick stem with almost no cap. Fortunately, unlike other oyster mushrooms, the stem of the king oyster remains soft and edible.

In commercial production, temperature, humidity, and CO_2 levels are kept at specific levels to affect the shape of the mushrooms. This can be a tricky and labor-intensive mushroom to cultivate, and each grower has his or her own specific growing medium mix that is often a tightly held secret. Premixed growing media are available from specialist retailers that generally produce high yields. We recommend growing king oyster mushroom indoors. For further information, see the chapter entitled "Cultivating Mushrooms Indoors" (p. 56).

Sheathed Woodtuft (*Kuehneromyces mutabilis*), Nameko (*Pholiota nameko*)

For many mushroom lovers, the sheathed woodtuft, native to Europe, is one of the most delicious. It grows in clumps, with stems that are fused at the base. The sheathed woodtuft has a pleasant, characteristic aroma. If found in the wild, be absolutely certain you've correctly identified it, as there are poisonous lookalikes.

Sheathed woodtuft can spend a long time on one piece of wood, and mushrooms often

Italian oyster mushroom

Sheathed woodtuft

Brown-gilled woodlover or conifer tuft

emerge from the soil surrounding logs it has colonized. Interestingly, slugs are uninterested in these mushrooms and leave them alone.

Brown-Gilled Woodlover or Conifer Tuft (*Hypholoma capnoides*)

The brown-gilled woodlover is an autumn mushroom, growing from early autumn to early winter, sometimes also in spring. Its fruit bodies grow in clumps, sometimes directly out of the soil from deadwood. The color of its cap can vary from pale yellow to brownish yellow to ocher. Its gills also have characteristic coloration, ranging from smoky gray to grayish-violet in fully grown mushrooms. Also known as conifer tuft, it is grown not only because of its mild aroma, but also because its preferred growing medium is conifer wood. In the wild, it can easily be mistaken for sulphur tuft, which is bitter-tasting and poisonous.

Enokitake or Enoki (*Flammulina velutipes*)

Enokitake is a mushroom that mainly fruits in winter and is thus a wonderful complement to other mushrooms in the garden. It is yellow to honey-brown in color and often grows in clumps. In deciduous forests, it loves to colonize willow stumps. In wet conditions, its surface can become slimy. In Japan, this mushroom is grown in bottles. Unlike European strains, Japanese strains have white caps. Whether in soups or sautéed, enokitake is full-flavored and has a pleasant consistency.

King Stropharia, Wine Cap, or Garden Giant (*Stropharia rugoso-annulata*)

King stropharia is found in the wild in Europe as well as in parts of North America and Japan. Its cap is typically reddish brown in color and its striking stem is thick and robust all the way down to its base. Wine cap or garden giant are other common names for this mushroom. Cultivation is best on unfermented straw bales that are in contact with the ground, as mycelial soil contact is crucial for mushroom formation.

Its mushrooms grow hidden in the bales and often do not emerge until fully ripe. They should be harvested by twisting away from the bales and then removing whatever part of the stem remains that would otherwise attract slugs and fungus gnats. King stropharia will add flair to your cooking and should be harvested when still small.

Sheathed Woodtuft, Nameko

Recommended growing medium: wood	Birch, beech, oak, maple, willow, alder (the first three are also appropriate for nameko)
Log diameter	• sheathed woodtuft is typically grown on logs with a diameter of 10–14 in (25–35 cm) • inoculate logs 3 feet (1 m) in length
Average colonization time	• 1 to 2 years (grain spawn) • we recommend auger or plug spawn method • after colonization: cut logs in thirds (see Oyster Mushrooms, p. 45)
Harvest	Spring and autumn (nameko only in autumn)
Unique characteristics	• bury logs • long colonization time • these mushrooms do not attract slugs • mycelium-soil contact must be well-established before mushrooms form
Culinary profile	• sheathed woodtuft: excellent in soups; only use caps • nameko: full aroma unfolds with sautéing and is great in soups

Brown-Gilled Woodlover or Conifer Tuft

Recommended growing medium: wood	Spruce
Log diameter	• this mushroom is typically grown on logs with a diameter of 10–14 in (25–35 cm) • inoculate logs 3 feet (1 m) in length • plug spawn (spruce dowels) or auger method
Average colonization time	• 1 year (plug spawn and auger method) • after colonization: cut logs in thirds and bury
Harvest	Early autumn to early winter, sometimes also in spring
Unique characteristics	• bury logs • conifer wood dries quickly, pay close attention to wetness of wood • only use completely healthy wood (e.g., free of infestation from bark beetles) • mycelium-soil contact must be well-established before mushrooms form
Culinary profile	• mild, slightly nutty flavor • use caps only; can be mixed with salt in dried, powdered form

Enokitake

Recommended growing medium: wood	Willow
Log diameter	• this mushroom is typically grown on logs with a diameter of 8–14 in (20–35 cm) • inoculate logs 3 feet (1 m) in length • grain spawn can be used
Average colonization time	• 1 to 2 years • after colonization: cut logs in thirds and bury with moss atop the cut face
Harvest	Mid autumn through early spring
Unique characteristics	• bury logs • fruits from mid autumn and throughout winter
Culinary profile	• pleasant forest mushroom aroma • only caps are processed

King Stropharia (Garden Giant)

Recommended growing medium: straw	Small bales from high-pressure balers are best
Environmental Requirements	• straw must be hydrated before inoculation (no need for fermentation) • cultures can be established in the garden and require soil contact
Harvest	• spring to autumn, depending upon time of establishment • several flushes of mushrooms over course of one year
Unique characteristics	• harvest before mushrooms grow too large • require protection from slugs • cultivation originated in Europe
Culinary profile	• excellent on the grill • cook thoroughly • stem can usually also be processed

Nameko in various stages of development

King stropharia

Enokitake

Wood ear or jelly ear fungus

Lion's mane

Lion's Mane (*Hericium erinaceus*)

Lion's mane is an excellent edible mushroom. Further, it is a cherished medicinal mushroom and used in Traditional Chinese Medicine (TCM) to treat stomach and intestinal ailments. It is found only rarely in the wild. Its white, icicle-like fruit body makes it a fascinating mushroom to cultivate.

Strains of this mushroom are native to North America, Europe, and Asia. Lion's mane can be grown on logs as well as on a sawdust growing medium. On sawdust, fruit bodies often fuse with one another, forming one giant mushroom. It should not be left on its growing medium for too long (its long spines should still be white) and should be processed immediately after harvesting. Fruit bodies should be cut carefully to avoid bruising the flesh. For mushroom connoisseurs, this one is a must.

Wood Ear, Jelly Ear, or Jew's Ear Fungus (*Auricularia auricula-judae*)

This mushroom has been cultivated as an edible for at least 1,500 years and is a staple in Asia. The fruit body, which resembles an external ear, has historically been called Jew's Ear. In the wild, it grows all year long on weakened deciduous trees, particularly on elder, beech, and willow. Wood ear can also be used as a medicinal to improve blood circulation.

Lion's Mane

Recommended growing media: logs, sawdust substrate	Logs	Sawdust
	• beech, birch, maple, oak • auger method recommended	Commercial sawdust growing medium
Needs for successful cultivation	• 8–12 in (20–30 cm) diameter logs • can be left as meter-long logs	• growing medium retains moisture on its own • enclosing during fruiting can help raise humidity in air
Harvest	Summer, autumn	Several flushes of mushrooms over period of 3 to 4 weeks
Unique characteristics	Harvest fruit bodies in a timely fashion	Bag of growing medium to remain unopened during mushroom flushes. Poke 3 to 5 holes in bag (this keeps humidity high in bag and minimizes contamination risk)
Culinary profile	• diverse: fine mushroom flavor, slightly zesty • when cooked, reminiscent of lobster • a true delicacy	

Wood Ear

Recommended growing medium: logs	Elder, poplar, willow
Needs for successful cultivation	• this mushroom is typically grown on logs 3–6 in (8–15 cm) in diameter • inoculate meter-long logs
Harvest	• Late summer to early spring (most frequent mushroom flushes), usually fruits throughout year
Unique characteristics	Once dried, mushrooms are good for at least a year in tightly closed jars
Culinary profile	• can be used to thicken sauces • firm flesh; often used in soups

Reishi (*Ganoderma lucidum*)

In the chapter entitled "Protected Environments for All Seasons" (p. 85), you will find a detailed description of reishi.

Reishi

Recommended growing media: logs, commercial sawdust growing medium	Beech, hornbeam
Log diameter	Reishi is typically grown on 3-foot-long (1 m) logs with a diameter of 6–12 in (15–30 cm)
Average colonization time	Grain spawn: auger method—1 year
Harvest	• mid to late summer • produces mushrooms for several years
Unique characteristics	• after colonization: cut logs in thirds and place in pots • requires protected atmosphere
Culinary profile	• slightly bitter and earthy • only for use in teas or as powder or extract (see Recipes and Processing Edible Mushrooms, p. 118)

Reishi grows well on beechwood logs

Oyster mushrooms

2. Cultivating Mushrooms Indoors

Mushrooms on Straw Pellets— Practical and High Yielding

A newer, extremely simple method for mushroom cultivation that uses straw pellets is becoming popular. These can be found in farm stores being sold as bedding for horses and are made of shredded, pressed straw. If possible, find pellets made from organically grown straw.

A further important component is an appropriate grain mushroom spawn. In general, this is a growing method for warmer months. Thanks to acids produced when undergoing a process of fermentation, the straw becomes somewhat solubilized and is turned into an optimal growing medium for mushrooms. As this fermentation process can result in fairly unpleasant odors (think silage), ferment straw in a covered bucket. Temperatures in a range of 59–68° F (15–20° C) are ideal for establishing a culture.

This is an ideal method for establishing mushroom beds. A rectangular box or an empty cold frame can be used to this end. The frame helps in maintaining a moist microclimate (depending on temperature it may be necessary to cover with a cloche—this tends to also help against slugs). Buckets or large flower pots (with holes) can also be used as mushroom growing containers. There are hardly any limits to creativity here. Follow the practical considerations, however.

Substrate Preparation

Straw pellets must be soaked in water for about seven days, after which the fermentation process will have sufficiently run its course. Remove the outer dark brown layer (½–1 in [1–2 cm]) that will have formed. At this point you will need to have your grain spawn, about 2 quarts (2 L) of which will inoculate 11 lb (5 kg) of straw pellets. The pellets will greatly expand in volume. 5½ lb (2½ kg) of them will require about 2 gallons (8 L) of warm water, so be sure to use a container that can accommodate this. Fermenting outdoors reduces your exposure to the odors produced; when fermenting indoors, cover the container to minimize odors.

After seven days of soaking, remove all excess water. Wet straw stains clothes and skin; wearing disposable gloves and work clothes will help you keep clean. Water will drain through the holes of the final container (flower pot, bucket, etc.), so some kind of containing vessel will be necessary. Any surplus water here should also be removed.

Inoculation

Your mushroom spawn (golden oyster, blue oyster, Italian oyster, late oyster, pink oyster) can now be evenly mixed into the mass of straw and placed in its final container—a good time to use disposable gloves. Compact the inoculated growing medium somewhat.

Closing up your container with plastic wrap helps achieve optimal growing conditions for the mycelium you are cultivating by encouraging a moist microclimate. Poke a few holes in it so that fresh air can still enter the container. Containers should not be filled all the way to the top with inoculated substrate.

Colonization Phase

Store your container indoors in the basement or cellar, or outside in the garden or greenhouse

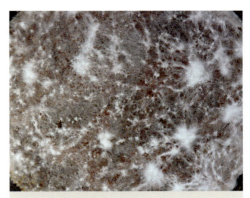

Mycelial growth on straw pellets

if it's more convenient. It is normal for water droplets to form on the inside of the plastic wrap—they are harmless. After several days, white mycelium will appear on the surface of the growing medium. It will take about four weeks for mycelium to completely colonize the straw. Once the first mushrooms start to appear, remove the plastic wrap so they receive plenty of fresh air and light. Give your culture one or two sprays of water per day during fruiting.

Harvest

Once your mushrooms have grown to full size, carefully cut them free of the growing medium. Oyster mushrooms grow together in clumps—if you only harvest the largest mushroom of the cluster, the surrounding, smaller mushrooms will die. For this reason, harvest entire clusters at once. Mushrooms often grow out of the holes in your container.

Do not allow the mushrooms to grow too large, as they may drop spores. Your culture should not be allowed to dry out as long as the growing medium is still producing, but overwatering is also problematic. Little by little the mushroom culture will take up nutrients extracted from the straw substrate. A sign of this decomposition process is the shriveling up

Ripe mushrooms ready for harvesting

of the growing medium. Mushrooms will come in several flushes.

Dark-winged fungus gnats, attracted to the smell of the mushrooms, can become problematic. These insects do not directly attack the mushrooms, but feed instead on the growing medium. Placing yellow panels in the immediate proximity of your culture can help lure them away. Spent substrate can be composted or used as mulch or fertilizer in the garden.

Mushroom Kits—The Perfect Way to Cultivate in Wintertime

Most plants and fungi are dormant over winter. For those who do not wish to go without fresh delicacies from the fungal kingdom in this season, mushroom kits are the answer.

Mushroom kits are excellent for starting out with mushroom growing. Furthermore, it is fun to observe mushrooms as they grow before your eyes. Mushroom kits are containers with pre-inoculated and colonized growing medium. Basements, cellars, or even living rooms are appropriate spaces for growing kits. All that needs to be done is to cut open the bag and keep it moist.

There is a large diversity of mushrooms available in kits: button mushrooms, shiitake, oyster mushrooms, and other less common edible and medicinal mushrooms (like king oyster, lion's mane, reishi, etc.). By buying growing medium that has already been inoculated and colonized, you skip these phases and can harvest that much sooner.

Requirements for Successful Cultivation

- Air should not be too dry
- Keep moist with a spray bottle (twice per day is usually sufficient to maintain optimal moisture levels)

An Overview of Mushrooms Commonly Available in Growing Kits

Type	Culinary Profile	Ideal Temperature Range
King oyster (Pleurotus eryngii)	Full-flavored, firm flesh	50–64° F (10–18° C)
Lion's mane (Hericium erinaceus)	Very aromatic, tender flesh	50–75° F (10–24° C)
Poplar mushroom or pioppino (Agrocybe aegerita)	Strong mushroom aroma, nutty flavor	64–86° F (18–30° C)
Oyster mushroom (Pleurotus)	Tender, aromatic	43–75° F (6–24° C)
Shiitake (Lentinula edodes)	Characteristic aroma, firm flesh	57–72° F (14–22° C)
Reishi (Ganoderma lucidum)	Slightly bitter/earthy, for use as tea, powder, or extract only	68–81° F (20–27° C)
Button mushroom (Agaricus bisporus)	Classic mushroom flavor, can also be eaten raw	54–64° F (12–18° C)
Buna shimeji (Hypsizygus tessulatus)	Tough flesh, exotic appearance	55–64° F (13–18° C)
Almond mushroom (Agaricus blazei Murrill)	Unmistakable aroma	64–75° F (18–24° C)

- The smaller the mushrooms, the higher the humidity in the air they will need
- Kits need access to fresh air and sufficient light
- Do not allow the kit to become waterlogged
- Store the kit within the optimal temperature range for the mushroom being cultivated (normally indicated in the text on or included with the kit)

For more information about each kind of mushroom, see individual entries in the Mushroom Portraits, p. 42 and p. 64.

Cultivation Guide

To begin, simply cut open the bag. Mushrooms grow in the direction where they get the most oxygen. If you notice that mushrooms are not growing, remove the bag completely and set the block of inoculated substrate in an appropriately sized plastic container (*see photo opposite*). Two pieces of wood at the bottom allow for excess water to drain out of the growing medium and promote better air circulation.

For lion's mane and oyster mushrooms, cut more holes in the mushroom kit bag. For some species, it can be good to scratch the surface of the substrate (while still in the sealed bag) and try different locations if mushrooms do not develop. Initiating temperature changes (from warm to cold or vice versa) can also help shock the culture into fruiting.

Magnificent king oyster mushrooms

Harvest

As soon as the bag has been cut open, ripe, ready-to-harvest mushrooms will appear within one to three weeks. Mushrooms should be harvested when their caps are still oriented slightly downward. Don't wait too long to harvest, as overripe mushrooms will begin to disburse spores into the environs. If only a few large mushrooms appear, it could mean that it is too warm or not moist enough. Sometimes it takes longer for the first mushrooms to appear—do your best to remain patient, as even under the best conditions, natural growth processes need their time. The window of time for harvesting can range from a few weeks (button mushrooms) to a few months (reishi). When the growing medium has all shriveled up and it nearly ceases taking on water, allow it to dry out. After several days, submerge it completely (weigh it down) in cold water for two to three hours. Place it back in the container and wait for more mushrooms.

A Note About Shiitake Mushroom Kits

As with growing shiitake on logs, shiitake growing kits also require a rest period between mushroom flushes. The kit can be reactivated by submerging the dried block of substrate in water. To speed this along, you can drill holes in the block (using, for example, a screwdriver or a knitting needle). After submersion, the block will again be heavy and should be moistened daily. Use cold water for submerging. Soon mushrooms will again appear.

OUR TIP: *Should mold appear, scratch it away and treat affected sites with vinegar or salt (dab with paper towel, but only where mold appeared).*

King oyster mushrooms

Mushrooms from Compost

Button Mushroom Growing Kit

Button mushrooms (champignons) are extremely tasty mushrooms that are widely enjoyed and appreciated. Growing button mushrooms, however, is not for beginners. Practically speaking, cultivating them from spore to fruiting substrate is only financially viable when done at a very large scale and sold wholesale. For those who are eager to learn more, though, we provide here a short summary of the most important steps in button mushroom cultivation, from A to Z (see Making Compost for Button Mushrooms, p. 63).

For the hobby grower, who would like some occasional mushrooms and enjoys watching them grow, procuring a growing kit is likely to make the most sense. In the winter, ideal growing temperatures are to be found indoors. At the same time, be aware that only a few mushrooms will grow above 68° F (20° C). There are normally two varieties available commercially: white and brown. The brown variety is more intense in flavor.

How to Use a Button Mushroom Growing Kit

Most commercial button mushroom growing kits include a growing medium (compost) pre-colonized by mycelium and a quantity of peat moss called casing, which you are to loosely cover the compost with by hand to a depth of about 1¼ in (3 cm). Store in a cool place (ideally 54–46° F [12–8° C]). Do not compact the casing; allow about four weeks for the mycelium to grow into it. Keep a close eye on the moisture levels; if the surface becomes dry, moisten with a spray bottle.

The Casing and Optimal Moisture Levels

The fungus requires a covering layer that the mycelium can grow into. Without this so-called casing, the fungus will not be able to produce fruit bodies. In growing kits, peat moss, with some natural additives like lime, is normally used for casing. It can store great amounts of water, which helps create ideal growing conditions for your mushroom culture. If the casing has been packaged moist, it will need no additional moisture for the first few weeks after opening.

The mycelium will grow into about the lower inch (2 cm) of the casing. Within about two weeks, a fine net of mycelium will become visible. Mycelium should not grow up to the surface of the casing, however. Should this occur, scratch the upper layer open with a clean fork and return the kit to a cool place.

Only spray with water when the substrate has become dry. Too much water can kill the mycelium.

Moisture Content

It may be the case that there is too little humidity where the mushroom kit has been situated. If so, you can provide the kit with a mini-greenhouse or cover with plastic wrap. Once the first mushrooms start to emerge, they will need more oxygen, so remove any such covering at this time.

Harvest

Depending upon how quickly mycelial growth proceeds and how close to ideal the humidity and temperature are, the first mushrooms should appear within three to four weeks. Emerging mushrooms need plenty of fresh air (the smaller the mushrooms, the more oxygen they require). Harvest mushrooms when the cap is only slightly curved inward. Carefully twist—do not cut—mushrooms away from the growing medium.

Mushrooms will store for four to six days in the refrigerator. After the first flush of mushrooms, the growing medium will need time to recover, during which the casing can be allowed to dry out. After two weeks you may rehydrate. Harvesting can cause the growing medium to be unevenly distributed. Scratching the surface open and redistributing the casing with a clean fork, then watering, will "reset" the culture to flush anew.

Button mushrooms ready for harvest: here the various layers of the growing medium can be easily recognized

At this point you will be able to observe a blending of the various substrate layers. The kit should be good for one to three flushes of mushrooms before being exhausted of nutrients required by the fungus for fruiting. If you have a compost pile, dispose of it there. With luck, mycelium will spread into the pile and it will produce more mushrooms.

Making Compost for Button Mushrooms

The main ingredients for button mushroom compost are horse manure, dried chicken manure, and gypsum. These are mixed in a pile (at least 53 ft^3 [1.5 m^3]) and allowed to ferment as preparation for inoculation. You will need at least 1,323 lb (600 kg) of horse manure, 132 lb (60 kg) of dried chicken manure, and 33 lb (15 kg) of gypsum. The most labor-intensive part of the whole process is the layering and turning of the compost. The pile is to be built layer by layer, watering every 1 ft (30 cm), until it has reached a height of 5 ft (1.5 m). The final layer should be tamped down somewhat. If everything works as it should, fermentation should begin within a few days and the temperature of the pile rises to 176° F (80° C).

This process is indispensable for inoculating with button mushroom mycelium. In order for oxygen to reach the innards of the pile, it must be turned occasionally, with the first turning to be done about a week after establishment of the pile. Every layer (about every 1 ft [30 cm]) needs to be watered again when turning. The pile should be turned every week, three times in total, after which the fermentation process will have run its course. In commercial production, the compost is then sterilized before inoculation, but for hobby growers, there will likely be no means by which to do this and fermentation will have to suffice.

Inoculating with Button Mushrooom Spawn

For inoculation, raised beds about 2½ ft (80 cm) wide and 8–12 in (20–30 cm) deep are established on the ground. Once the compost has cooled to about 77° F (25° C), spawn is to be mixed in. About 6½ cups (1.5 L) of spawn are required for each square meter (about 11 ft^2).

Colonization Phase

The ideal temperature range for rapid colonization is 68–75° F (20–24° C), with fairly high humidity. Within two to three weeks, the mycelium will have established itself in the bed. After colonization:

- Cover the compost with 2 in (5 cm) of casing. Whatever is used for casing, it must be able to hold lots of water and have a pH in the range of 6.5 to 7.5. If making your own casing mix, peat moss can be mixed with equal parts garden soil. The pH value can then be adjusted by adding lime. Before covering beds with this mix, it must be watered— sufficiently such that balls can be formed, but not so wet that water then drips out.
- Ten days after covering the beds with casing, mycelial growth will start to become visible. The surface should then be scratched with a steel brush, which spreads the mycelium across the entire surface. The culture requires plenty of fresh air at this point. Light is not necessary.
- Fruiting phase: The ideal temperature for fruiting is 61–64° F (16–18° C), which is cooler than for colonization. Little by little, the beginnings of mushrooms (pinheads) form. Mushrooms are extremely sensitive in this phase, during which the growing medium should never dry out. Watering should only be done with a fine-spraying watering can to protect the surface from becoming slimy.

Harvest

Mushrooms are ripe enough to harvest when the cap has separated from the stem. Mushrooms will come in several flushes. After each harvest, the growing medium should be moistened anew, as the mushrooms remove much water from it.

Possibilities for growing sites and appropriate containers:
- shady part of garden (spring, autumn)
- tunnels
- a humid basement or cellar with sufficient access to fresh air (should be easy to clean well)
- raised beds, *hügelkultur* beds (not unlike beds for asparagus)
- long wooden crates

When cultivating in a basement or cellar, the temperature should stay below 68° F (20° C). Air circulation via a window or fan is recommended.

OUR TIP: *This is a mushroom for experienced hobbyists and experts. Do not underestimate how much work they are. Still, do not be discouraged by early failures. See the bibliography for specialist literature if you need help with more specific problems (see Appendix, p. 139).*

Reasons for crop failure:
- contamination by bacteria, viruses, other fungi, or insects
- mold (treat contaminated sites early with salt)
- incomplete fermentation
- air temperature too high
- CO_2 levels too high
- incorrect compost and/or casing mix

Additional Mushroom Portraits

Portobello mushroom

Photo: Moritz Wildenauer

Button Mushrooms (*Agaricus bisporus*)

Recommended growing medium	Fermented compost (consisting primarily of horse manure, dried chicken manure, and gypsum) and casing
Environmental needs	• requires no light • temperature: colonization phase 68–75° F (20–24° C), fruiting phase 61–64° F (16–18° C)
Harvest	Varies by site: in garden, spring or autumn; in basement or cellar throughout year
Unique characteristics	This fungus is considered to be a secondary decomposer (grows only on "pre-digested," fermented substrates)
Culinary profile	• firm to the bite • delicate mushroom aroma • can be eaten raw in salads • see Recipes and Processing Mushrooms, p. 118

CULTIVATING MUSHROOMS INDOORS

Almond Mushrooms (*Agaricus blazei Murrill*)

Recommended growing medium	Fermented compost or straw/compost
Environmental needs	Temperature: colonization phase 70–81° F (21–27° C), fruiting phase 64–75°F (18–24° C)
Harvest	• throughout the year in basement, cellar, or house • carefully twist away from growing medium when harvesting
Unique characteristics	• excellent edible and medicinal mushroom
Culinary profile	Can be used anywhere mushrooms are called for (prepare like button mushrooms)

Buna Shimeji (*Hypsizygus tessulatus*)

Recommended growing medium	Growing kit with beechwood chips and sawdust
Environmental needs	Temperature: colonization phase 70–75° F (21–24° C), fruiting phase 55–64° F (13–18° C)
Harvest	• protected environments (mini-greenhouse, basement, cellar) make a big difference in success of crop • cut away from substrate
Unique characteristics	Wild appearance (spotted cap or snow-white cap coloration)
Culinary profile	• good edible mushroom • goes well with garden greens • flavor unfolds when covered with butter

Photo: Walter Haidvogl

Almond mushrooms

Buna Shimeji

Pioppino emerges from its growing medium

Poplar mushrooms or pioppino (*Agrocybe aegerita*)	
Recommended growing medium	Growing kit with beechwood chips and sawdust
Environmental needs	• colonization phase 70–81° F (21–27° C), fruiting phase 55–64° F (13–18° C) • cultivate in closed spaces
Harvest	• protected environments (mini-greenhouse, basement, cellar) make a big difference in success of crop • cut away from substrate
Unique characteristics	Beautiful brown cap coloration
Culinary profile	• delicious edible mushroom • adds a decorative element to meals

3. Cultivating Mushrooms in Woodlands and Fields

Cultivating Truffles—The Secrets of Mycorrhizal Fungi

The truffle is one of the most expensive and sumptuous edible mushrooms of all. Its flavor is compelling and so is its mysterious symbiotic relationship with plants. Truffles are very specific in their preferences for soil type, temperature, climate, and which tree or shrub species they will partner with for symbiosis.

Fungi that form such a symbiotic relationship with plants are called mycorrhizal fungi. Both plant and fungus benefit by exchanging substances with one another. The fungus receives carbohydrates from the plant, and in exchange the plant receives macronutrients (nitrogen, phosphorous, potassium) and trace elements. Through mutual water exchange, both symbiotic partners can optimize growth. The fungus extends the reach of the plant's roots in effect with its hyphae, which access even the smallest pores in the soil. The truffle is a so-called ectomycorrhiza, along with many other mushrooms that can be found wild in the forest, such as chanterelles and porcinis.

The processes involved in these symbiotic relationships have been poorly understood and are nearly impossible to imitate in cultivation. However, thanks to extensive research into the lifecycle of the truffle, it is now possible to cultivate them under conditions that closely resemble those found in nature.

Commercially available truffles will either have been gathered in the wild or cultivated at truffle farms that are able to closely simulate the natural conditions that truffles grow in. Not long ago, the first successful truffle-growing operation in central Europe was established in Austria by an innovative company called Trüffelgarten,

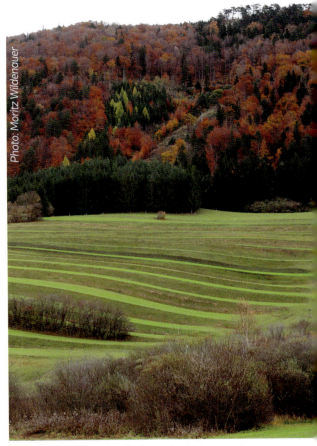

Photo: Moritz Wildenauer

under the leadership of Alexander Urban and Tony Pla. They founded the company in 2003 with the goal of establishing truffle cultivation in central Europe. Through their research, the two founders hope to make their contribution to biodiversity and support clients wishing to embark on truffle cultivation with the latest understanding and knowledge.

Trüffelgarten offers high-quality truffle-inoculated plant stock and also consultation on truffle cultivation. The company also offers consultation on siting considerations and hunting for truffles. Identifying truffle finds and the identification of ectomycorrhizas in general also fall into their scope of activities.

Ronald Vogl and Tony Pla with dog Titoune on the hunt for burgundy truffles. Ronald Vogl is the first truffle grower in Austria and indeed the whole of central Europe. His plantation (looked after by Trüffelgarten) is located in the province of Lower Austria, near Herzogenburg.

Important Points in Truffle Cultivation

Timing Your Planting

A tree sapling inoculated with truffle is best planted in mid to late autumn in the garden. This allows it to establish itself before winter, allowing for good spring growth the following year. As the saplings have been grown in pots, they can also be planted in spring, though this is more difficult because they will need more water and protection from the sun.

As already implied, truffles cannot simply grow in any soil or climate zone. If you have purchased multiple truffle saplings, plant spacing will vary by variety (note at time of purchase). If your truffle-inoculated plants are intended to be a hedge, they can normally be planted much closer than for regular hedging plants. Appropriate spacing should also be observed with existing trees and shrubs, as truffle-inoculated plants may outcompete non-inoculated plants living nearby.

Competition with Other Plants

You will also need to make sure that the truffle mycelium you are introducing will not be able to simply jump onto an existing tree or shrub. In soils where truffles are not native, it is difficult to maintain and propagate truffle mycelium. For this reason, it is recommended to grow as far away from forests as possible to avoid contamination from and, ultimately, being outcompeted by other ectomycorrhizae. Fruit trees, however, do not represent a threat to truffles, as they are surrounded by other kinds of fungi.

Climatic Conditions

The black truffle can be found all across Europe (from Italy to southern Sweden), so its cultivation is possible, at least in central Europe. In Austria, these truffles have been found at confirmed elevations of 1,804 ft (550 m). In dry periods, watering may be necessary.

The Périgord truffle grows in the wine-growing regions of southwest Europe and is not found in central Europe. Cultivation in central Europe is still in its infancy, though there have been some successes.

Soil Conditions

Truffles do best in lime-rich, well-draining soils. Soils that retain too much water and can become waterlogged are not appropriate. Where soils are low on lime, applying lime to them can be successful. Soil testing is crucial if larger plantations are planned, to determine your chances of success. The Austrian company Trüffelgarten offers a cooperative model, wherein site conditions are tested and analyzed.

Yield

Yields vary dramatically, and the first harvest doesn't even arrive until three to six years after planting. Each tree will produce around 5–10 oz (150–300 g) of truffles on average. Yet this is still dependent on several factors. In good years, several kilograms may be harvested. Per hectare, yield can vary from 13–26 lb all the way up to 130 lb (13–30 kg up to 150 kg per acre).

Inoculation Success for Producers

When a truffle sapling comes from a "truffle farm," the producer can guarantee colonization of the plant stock. Continued success of the truffle culture is dependent on many factors. Since successful cultivation in our native central Europe is still a fairly new phenomenon, it is impossible to guarantee success, even from "guaranteed" plant stock. The best one can do as a grower in an area without proven truffle success is to provide all the necessary care for the culture, establish the necessary conditions, and hope for the best.

In the Austrian province of Lower Austria, the company Trüffelgarten offers a service wherein they establish your truffle planting with you (trees, fences, irrigation, soil preparation, etc. for 15,000–22,000 euro [about 18,500–27,000 dollars] per hectare). This offers growers an opportunity to enter the potentially lucrative yet risky truffle market.

For those who would be happy to meet their own truffle needs with a tree or two in the garden, plant stock from Trüffelgarten offers a great regional alternative to expensive truffles from the store.

Hunting Truffles and Harvest Timing

The important question remains, when will you be able to find truffles in your own garden? You may have heard of using pigs or dogs for finding truffles, but they will not be necessary if you only have a few plants in your garden.

Burgundy truffle

Périgord truffle

Bianchetto truffle

A successful harvest of burgundy truffles

Sparse grass can indicate the subterranean growth of truffles (not unlike underneath a walnut tree). Another method, which requires sharp vision, is to search for truffle flies (*Suillia tuberiperda*). These insects swarm directly over truffles hidden in the ground. Shoo them away with a stick and follow them wherever they go next to find the next ripe truffle. Truffles may also protrude from the ground, so you can simply look for them as well.

Harvest time for the Périgord truffle is in winter, for the burgundy truffle summer to late autumn. The exact timing depends on the prevailing weather conditions and varies from year to year.

Tree Species for Truffle Cultivation

There are several factors to be considered when selecting tree and truffle species for cultivation. The inoculated tree must grow in the conditions that prevail at your site, so the type of tree is perhaps the most important decision.

- **Hornbeam (*Carpinus betulus*):** Hornbeam is fairly flexible and can be easy to integrate into the garden, for example in the form of a hedge. It also responds well to pruning and partners well with truffles.
- **Turkish filbert (*Corylus colurna*):** A very drought-resistant species. Good for drier sites, and you can harvest nuts in addition to truffles.
- **Oaks (*Quercus cerris, Quercus robur, Quercus pubescens*):** Different species have their own site preferences. These range from dry, warm sites (*Q. pubescens*) to somewhat clayey soils in hotter areas (*Q. cerris*). The site should, however, not be so clayey as to cause waterlogging in the soil, as truffles do not do well in such conditions.
- **Atlas cedar (*Cedrus atlantica*):** This tree can grow to be up to 131 ft (40 m) tall. This cedar is best for areas with mild winters. It is fairly drought tolerant and can only be grown in areas with hard frosts when it can be protected. Very promising for truffle cultivation.
- **Austrian pine (*Pinus nigra* ssp. *austriaca*):** This hardy pine can be used to grow truffles from the eastern edge of the Alps to lower-lying regions. It needs a lot of light and a rocky, limestone- or dolomite-based, nutrient-poor soil.

Establishing a Bed for Mushrooms

Setting up a mushroom bed outdoors is quick and easy. The best time for inoculation is late spring, when frost is no longer expected. Mushroom spawn with a higher proportion of wood works best here, as grain spawn tends to attract mice and other pests that may adversely affect colonization.

There are many different materials that could be used to establish a mushroom bed. Straw, hardwood chips, and compost have all been used historically. The growing medium must in any case be appropriate to the type of mushroom being grown.

Preparing a Mushroom Bed

Required materials:

- 2½ quarts (2½ L) mushroom spawn on a wood-straw growing medium (with hardwood chips and sawdust)
- 2 straw bales or chopped, dry straw
- covering material (may be soil)

Alternative to straw:

- 15 quarts (15 L) hardwood chips (untreated)
- 10 quarts (10 L) coarse sawdust
- 10 quarts (10 L) fine sawdust
- covering material: bark mulch, straw, or soil
- spade and shovel

When growing shaggy mane, oyster mushroom, and king stropharia, work chopped straw into the bed. We have found a combination of hardwood chips and additional straw also yields good results.

For beds established in the garden or woodland, select a shady spot under trees or shrubs. As circumstances require, you may need to fine-tune the microclimate—if there is no shade, for example, set up awnings or create shade some other way. If you plan on setting up several beds next to each other, make sure there is sufficient space between them to allow for care and maintenance of the beds.

Inoculation

With a spade, remove soil to a depth of 4 in (10 cm) in a 2 ft² (60 cm²) section of ground. Alternatively, you can establish a raised bed by using wooden framework of the same dimensions. Cover the bottom of the bed with corrugated cardboard (not paperboard) as a base layer. Then spread about half of your dry straw or hardwood chips atop the cardboard and water well with a garden hose. After ten minutes it should be sufficiently wet and ready for

A small king stropharia mushroom on straw

inoculation. Break up the mushroom spawn while still in the bag, spread evenly over the surface of the bed, and cover with your remaining straw or wood chips. Water again for about ten minutes.

Colonization Phase

Check the moisture content of your bed once a week. As a rule of thumb, water whenever the bed is dry at a depth of just over 1 in (3 cm). While mycelium colonizes the bed, which takes four to six weeks, you may cover with cloches or row covers. The bed should also be covered to protect from long periods of rain. Under ideal conditions, mushrooms will flush the same year the bed was inoculated. In a bed with hardwood chips and straw, mushrooms will also come the following year. Colonization is complete when mycelium is visible all across the growing medium.

Caring for Your Culture

After the bed has been fully colonized, remove cloches or row covers. When growing king oyster mushroom, cover the bed with soil 2 in (5 cm) deep. Once the beginnings of mushrooms become visible, water daily. Slugs can be problematic here, so protect against them with slug-proof fencing. However you keep slugs at bay,

Courgette/zucchini and shaggy mane make good companions

never use poisons of any kind as these can be taken up by the mushrooms. If you've made a raised bed with a wooden frame, water-permeable cloches or row covers of unwoven polypropylene will protect your culture from slugs.

Before winter, to feed and protect your mushroom bed, cover it 2 in (5 cm) deep with hardwood chips. Once spring comes, resume your weekly checking for moisture content.

OUR TIP: *Mushrooms and vegetables can share the same vegetable bed. In early spring, you can inoculate a straw bed with oyster mushrooms and later plant this bed with seedlings. Simply push away the straw where each seedling is to be planted out, then close the straw back around the plant. The seedlings will need to be large enough to poke out of the straw. Mushrooms and vegetables have shown themselves to be extremely compatible with one another. Vegetables selected for such a polyculture should be fairly shade tolerant. Zucchini and other squash also pair well with mushrooms, as their large leaves create plentiful shade for mushrooms growing in the understory.*

Speciality: Shaggy Mane in Beds and Containers

In cultivating shaggy mane, there are several different techniques, each appropriate to different site conditions. If you happen to have the good fortune to live near a button mushroom grower, you can use uninoculated button mushroom compost as your growing medium. It will already have been sterilized and needs only to be inoculated with shaggy mane and placed in a clean place. Clean plastic buckets or flower pots make for good containers. Affix weed membrane or floating row cover from the garden to the top of the container to keep fungus gnats at bay.

Establishing a Polyculture

Another style of mushroom cultivation, recommended by Austrian mushroom grower Walter Haidvogl, is the direct inoculation of kitchen garden or vegetable beds. Zucchini growing near a compost pile can share space with shaggy mane. Growing medium colonized by shaggy mane mycelium can simply be added to the soil at the base of the plant. Just dig a small hole where the plant is already throwing shade and toss in a block of colonized substrate or spawn. Then cover with 3 in (7 cm) of compost.

ADVICE FROM WALTER HAIDVOGL: *A good casing layer allows air to reach the growing medium below and has a friable structure, such that it does not simply turn into mud when watered. Two to three parts garden soil mixed with one part peat moss is a good basic casing mix. It should be watered before being applied to the growing medium.*

Colonization Phase

The polyculture helps optimize moisture levels; the large leaves of the plant throw shade upon the mushrooms below. Depending upon temperatures and time of year, the mycelium has

Shaggy mane

Shaggy Mane (*Coprinus comatus*)

Recommended growing medium	• fermented button mushroom compost • straw or polyculture with plants
Environmental needs	• temperature: colonization phase 70–81° F (21–27° C), fruiting phase 64–75° F (18–24° C) • fruiting triggered by low temperature
Harvest	• varies by site: in garden (mid spring to mid summer), in basement or cellar throughout the year • harvest in a timely manner, as mushrooms quickly dissolve
Unique characteristics	Although this mushroom is considered to be a secondary decomposer, it can also function as a primary decomposer and can thus thrive in straw-compost substrates as well as in polyculture with vegetables
Culinary profile	Delicious sautéed or breaded

Mushrooms for Inoculating Beds of Straw or Hardwood Chip-Straw

Mushroom	Notes
King stropharia (*Stropharia rugoso annulata*)	Straw base with soil casing
Shaggy mane (*Coprinus comatus*)	Straw-hardwood-chip-compost base with casing or on compost in vegetable bed
Oyster mushrooms (Italian, golden, elm)	• straw base • may be grown in polyculture with vegetables (lettuce, spinach, Swiss chard, peas, wild strawberries, zucchini, squash)

finished spreading and mushrooms have begun to grow. It is possible to have mushrooms from mid spring through mid summer.

Harvesting Mushrooms

Several mushroom flushes will occur in the year of inoculation. Several mushrooms may also appear the following year. It is important to harvest mushrooms before they become overripe as they break down quickly into ink.

If you have no garden, there are commercial growing kits for shaggy mane that yield quite well (see Mushroom Kits—The Perfect Way to Cultivate in Wintertime, p. 58).

Mycorestoration

With help, it is possible for some mushroom species to establish themselves in the soil. These can then help remove toxins from the soil and bring it back into balance. This is known as mycorestoration and represents a relatively new field of research. Much potential is seen in this field, and there have already been many successes in restoring contaminated soils using fungi in this way.

When fungi break down the primary structures of wood, they do this by employing enzymes. These fungi can use these same enzymes to neutralize, dissolve, and/or bioaccumulate heavy metals and toxins in the environment. By mixing a substrate colonized by mycelium into the soil along with an appropriate growing medium (e.g., moistened hardwood chips), the fungus can remove harmful substances from the ground. The mycelium spreads through the soil, removing toxins from the ground and bioaccumulating them as-is or in a broken-down form. When mushrooms form, these are to be harvested and disposed of. Mushrooms that have shown themselves to be appropriate for mycorestoration include sheathed woodtuft, enokitake, oyster mushroom, king stropharia, and reishi.

Inoculating Tree Stumps—Edible Mushrooms Encouraging Biological Succession

As previously stated, many mushroom species break down wood. The removal of tree stumps with the help of edible mushrooms has been practiced in Hungary for more than twenty years on poplar stumps. Poplar has been used there for paper and wood production. Harvesting these poplars left behind stumps that could not then be cleared. Inoculating them with oyster mushroom spawn helped clear stumps and yielded mushrooms.

An example in the United States are *Pholiota* mushrooms, which are grown on various deciduous trees. In pine forests, the popular edible and medicinal hen of the woods/maitake (*Grifola frondosa*) has long been grown successfully. Maitake grows not only on pine, but also on larch and spruce.

There is almost no end to the list of possible inoculation sites. In nearly every garden or forest, at the edge of every field, there are stumps that could be used to grow edible or medicinal mushrooms. The essential factor, however, is the timing of inoculation. The stump should have been created no longer than three to four months before inoculation. Waiting longer reduces chances of success. It is worth considering that the mycelium must colonize the entire underground root system and reckon with the tree's lingering biological defenses, so stumps can take a long time to colonize. We recommend the auger method for inoculation as being easiest.

Auger Method

With a cordless drill and an auger bit, drill holes 3–5 in (8–12 cm) deep every 2–3 in (5–7 cm), regularly over the entire face. We recommend a bit $5/8$–$13/16$ in (16–22 mm) in diameter. You can also drill and inoculate from the side.

Backfill the holes with crumbled grain spawn using a stick and seal. A funnel that fits in the boreholes may be useful here. Wooden furniture dowels can be used for sealing. The purpose of sealing is to protect against pests and excessive drying.

Additionally, the inoculated holes should have wax applied to them. To accomplish this, either brush with melted cheese wax or dunk the wooden dowels in wax before pounding in. Finally, cover the entire face with a piece of moss and affix with staples. This also helps prevent excessive drying.

OUR TIP: *Long dowel rods of beechwood can be found at hardware stores or farm stores. Cut these down to ¼–½ in (5–10 mm) in length and use to seal boreholes.*

Required materials for the auger method:

- grain spawn
- cordless drill with auger bit (bit diameter: $5/8$–$13/16$ in [16–22 mm])
- sticks for backfilling boreholes with spawn
- funnel
- hammer
- cheese wax or beeswax
- moss (to cover cut face of stump)
- stapler (to attach moss)

Holes can be bored into the cut face and also the side of the stump

Plug spawn can also be used. To maximize chances of success, many plugs should be used. It is important to then seal with wax (see Plug Spawn Method, p. 22).

If you intend to inoculate a stump that is not in your own forest or garden, be sure the owner is in agreement. In our experience, people are quite willing to cooperate and are happy to see another use come of the stump. If you take the time to explain the process, skepticism on their part often melts away. Be sure to mention that your spawn could not possibly "infect" standing, healthy trees.

Inoculating a willow stump

CULTIVATING MUSHROOMS IN WOODLANDS AND FIELDS   77

Spawn is inserted . . . *. . . and sealed with wooden dowels*

Required materials for inoculating a stump

Field margin in the Waldviertel region of Austria

Matching Mushroom Species with Tree Species

Our experiment: In the winter of 2013, an overgrown field margin not far from our our mushroom garden in Waldviertel, Austria, was thinned out. The field had become too shady—so poplars, cherries, willows, oaks, birches, hazels, and elders were all cut down. After securing permission from the farmer, we inoculated the stumps in mid to late spring, about three months after they were felled. We had a diverse array of mushroom spawn at our disposal (golden oyster, blue oyster, Italian oyster, late oyster, shiitake, sheathed woodtuft, enokitake). Since we had about twenty stumps of various diameters that we wanted to inoculate, we were able to select a mushroom species for each tree species.

Oyster mushrooms generally have strong, vital mycelium, so we inoculated the wild cherries with various oysters. For the poplars, we used shiitake and oysters. Willows and hazels got enokitake.

We had to wait a bit to inoculate the birches because quite a bit of sap accumulated on the cut face after felling, which allowed competing organisms to colonize. In waiting a few weeks, we were able to remove the upper layer of the stump and then inoculate with woodtuft or nameko.

For the elder shrubs, wood ear would be the appropriate mushroom. Because the stumps were now no longer in the shade, we covered them with moss to prevent them drying out. Limbs and leaves could have been used had no moss been available. The vegetation around the stumps is more able to grow now with more access to light, which will eventually provide for more shade and improve the microclimate for the stumps. The more spawn used, the better the chances for success. The first mushrooms may take two to three years to appear. Labeling trees with the inoculated mushroom species is crucial for not missing out on mushroom harvests. After several years, few people will remember which tree was inoculated with what spawn. In this way, in summer you can look out for golden oysters and in autumn you can keep your eyes peeled for blue oysters or sheathed woodtuft.

There are several factors that can influence success or failure. What you can control is the amount of spawn used, timing of inoculation (spring), and care as colonization proceeds. We also need Nature's cooperation, however: Fast-growing vegetation to establish a moist microclimate and sufficient rain are the most important elements.

A further example of successful stump inoculation comes from the southeast Austrian province of Burgenland. There, oak stumps were inoculated with shiitake and produced mushrooms for several years. This is especially worth mentioning, as this is not supposed to be possible according to the specialist literature.

Our conclusion is that inoculating stumps contributes to biodiversity and is a good alternative to conventional techniques for clearing stumps. You not only get delicious edible mushrooms, you also accelerate the decomposition process of the wood.

Late oyster mushrooms emerge from the moss and foliage later in the year of inoculation

Italian oyster mushrooms grown in a container

4. Container-Grown Mushrooms for Courtyards, Balconies, and Patios

When you think of mushrooms, you will probably imagine overgrown gardens or moist, mossy sites. This is only logical, as mushrooms thrive in moist environments.

But what if you don't have an appropriate garden, much less your own wood? A courtyard or a balcony can be made into a mushroom garden with a little effort. All you need is a bit of imagination and gardening know-how to start growing your own mushrooms in the middle of the city. In order for mushrooms to thrive on a balcony, patio, or in a courtyard, you'll need a companion planting of shade-providing vegetables or ornamental plants.

Criteria for an Appropriate Site
- wind protection
- shade or the ability to create shade
- sufficient space: you will need 9–18 ft² (1–2 m²)
- nearby tap or faucet

Setting Up a Mushroom Garden on the Balcony or Patio

The first step is to decide how much space can be set aside for this endeavour; 9–18 ft² (1–2 m²) provides sufficient space to do several combinations of different cultivation systems. For such a space, you should be able to have three buried mushroom logs and perhaps another two straw pellet cultures.

CONTAINER-GROWN MUSHROOMS FOR COURTYARDS, BALCONIES AND PATIOS

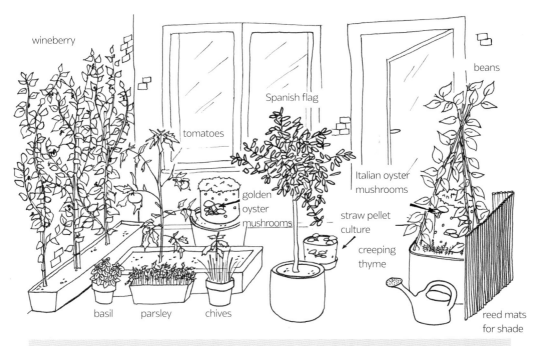

Balcony garden with mushroom cultures

Buried logs can be purchased either as pre-inoculated logs or you can inoculate them yourself. In the later case, there are several details to pay attention to. Mushrooms such as golden or blue oyster mushroom as well as sheathed woodtuft work well in pots. If you inoculate them yourself, the colonization phase will proceed a bit differently than as described for growing in the garden. Cut the 3-foot-long (1 m) log into thirds immediately and use the auger method to inoculate (see Auger Method, p. 24). Fill the boreholes with grain spawn and seal. Then immediately plant the three logs into an appropriate container.

Follow these instructions to fill the container with soil into which an inoculated log can be buried. Once planted, cover the top of the log with moss to prevent it drying out. The diameter of the pot should be large enough to accommodate your log, plus 3–6 in (7–15 cm) of space for soil. Having plenty of space between log and pot also allows you to add plants to provide shade and keep the log's microclimate moist. Avoid using terracotta pots, as the contents will be heavy even before you plant them. Water must be able to drain from the pot, so plastic pots with drainage holes and a saucer work well. Fill the bottom third with gravel or sand (for drainage) and the next third with garden soil. The soil should hold water well, so do not use just any potting soil. In order to optimize the conditions for a year-long colonization phase and the following harvest season, you will need to decide on some plants to live in the pot with the log for shade.

Tomato plants are excellent at providing shade for mushroom logs. The denser the growth of your wall of plants, the better the microclimate for your logs. An additional Spanish flag (*Ipomoea lobata*) plant keeps sun rays off the balcony. Another option would be to grow climbing vines like peas, beans, or cucumbers, using the log as a trellis. The mound-like growth of

Nasturtium as a shade-providing plant

creeping thyme can also grow in free spots in the pot as ground cover—although the pot will need to be fairly large in this case, as mycelium requires not only soil contact but also nutrients from the soil, which should not all have been scavenged by the companion plants before the mycelium reaches the soil.

This kind of combined mushroom/vegetable pot is not only a culinary delight, but also easy on the eyes as well. Decorative plants can be used instead of vegetables if desired. Plants like clematis, morning glory, cathedral bells, nasturtium, black-eyed Susan, ornamental grape, and other climbing plants all work well in this context. Because shade will primarily be needed in the second year (after a year-long colonization phase), you can wait until the second spring to plant your pot. As the mycelium colonizes the log, it should be placed in a shady site and be well-protected from wind. Another possibility would be to use tightly anchored awnings or reed mats to create shade where needed.

Check regularly for moisture content by observing the vitality of the plants, touching the soil and/or the surface of the moss. After watering, excess water should be emptied from the saucer. Make sure that the unique needs of different mushroom varieties are being met (see Mushroom Portraits, pp. 33–4 and pp. 42–54).

In winter, simply leave your mushroom pots outside. In the following spring, you can check to see if mycelium has made its way to the bottom of the log, which normally would have happened by now. If that characteristic white net is visible, you can be confident that your mushroom has successfully colonized the log. Don't give up, though, if you don't get any mushrooms in this first year after colonization—it could simply be that environmental conditions during what would have been a fruiting phase were not ideal. Simply continue to observe your log and have patience.

If you are only inoculating one 3-foot-long (1 m) log, you will need just under 2 quarts (2 L) of mushroom spawn. With leftover spawn, you can establish two pots' worth of straw pellet culture, which will also yield sooner than your log culture.

OUR TIP: *In the chapter entitled "Cultivating Mushrooms Indoors," p. 56, you can find a detailed description for establishing a mushroom culture on a straw pellet medium.*

Courtyards are great sites for growing mushrooms. The criteria and instructions described above are also valid here. As there is likely to be more space available in a courtyard compared to a balcony or patio, you will probably be able to inoculate more logs, as long as there is sufficient shade. You may also be able to grow a greater

Mushrooms for Balconies, Patios, and Courtyards

Mushroom	Harvest time	Notes	Companion plants*
Golden oyster, Italian oyster	Mid to late summer	mushrooms sprout from soil may need protection from slugs (see Slugs, p. 90)	• **Vegetables:** Tomato (*Solanum lycopersicum*), cucumber (*Cucumis sativus*), pole beans (*Phaseolus coccineus*), peas (*Pisum sativum*), zucchini (*Cucurbita pepo*), eggplant (*Solanum melongena*), wineberry/dewberry (*Rubus phoenicolasius*) • **Herbs and Edible Ornamentals:** Thyme (*Thymus*), shiso (*Perilla frutescens*), basil (*Ocimum* spp.), borage (*Borago officinalis*), mint (*Mentha*), nasturtium (*Tropaeolum majus*), mallow (*Malva sylvestris* var. *Mauritiana*) • **Ornamentals:** Clematis (*clematis*), morning glory (*Ipomoea*), cathedral bells (*Cobaea scandens*), black-eyed Susan (*Thunbergia alata*), ornamental grape (*Vitis vinifera* subsp. *sylvestris*)
Blue oyster	Mid to late spring; early to mid autumn		
Sheathed woodtuft	Mid to late spring; early to mid autumn		
Nameko	Early to mid autumn	Keep moist and shady in autumn months	
Shiitake	Early to mid summer; late summer to early autumn	• log must be submerged • grows only in vegetated courtyards	
Luminescent panellus	Summer months	• not edible • a unique design element • only for very dark sites • fruit bodies glow in the dark	

Observe recommended spacing

variety of mushrooms if you have a courtyard. If the courtyard is already well-vegetated and thus already has an appropriate microclimate, shiitake can be grown on logs, in which case a tub or other container for submerging logs will be needed. For paved courtyards, select mushrooms that can be grown in pots. Whatever specific cultivation techniques you decide on, you should, of course, select the shadiest area in the courtyard.

OUR TIP: *Houseplants positioned around your mushroom cultures can provide the shade required in summer months.*

DIY mushroom greenhouse in the garden

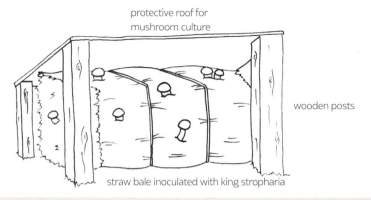

DIY shading solution for mushroom cultivation

5. "Protected" Environments for All Seasons

Mushrooms sometimes require protection from the sun, rain, or other less than ideal weather conditions. Some mushrooms (e.g., reishi, lion's mane, pink oyster) require more attention than others or involve specialized techniques, which suggest growing them in their own isolated, protected environment.

There are several ways to provide for such a protected environment:

- erecting a roof over your outdoor culture(s)
- building a mushroom greenhouse in the shade
- a smaller, indoor greenhouse (can be a simple plastic tunnel or more complicated technical equipment)

Advantages of "protected" environments:

- minimize fluctuations in temperature
- maintain high humidity
- minimize pests
- easier to control environmental conditions
- protect against unfavorable weather conditions
- rapid mushroom growth
- increased number of mushroom flushes

Erecting a Protective Roof in the Garden

A protective roof may be of great help in growing mushrooms during long periods of rain or if there is insufficient shade available in the garden. Prolonged rains soak straw cultures to such a degree that mycelial growth slows or mycelium may even die. Mushrooms being grown in beds on the ground can also benefit from a protective roof. If the purpose of the roof is only to provide shade, a wooden frame with reed mats or an awning should do the trick.

Wooden posts can be sunk into the ground and topped with a dual-layered reed mat (from a hardware or garden store) fastened with wire or string. The shade roof should extend beyond the mushroom culture. Select a height that allows space and access for harvesting.

Wherever or whenever protection from intense rains is required, it need not be difficult to protect your mushroom cultures. The simplest method is to cover the roof with a sheet of plastic and remove it when it's no longer needed. For a more permanent solution, you can build a more robust roof of polycarbonate twin-wall sheets. Topping with reed mats helps to maintain a more natural aesthetic.

OUR TIP: *When using a roof, check your cultures for moisture content more often, as they may dry out faster when protected from all rains.*

Small Indoor Greenhouses for Edible Mushrooms in Winter

There's nothing better than cultivating your own mushrooms in the winter. Since heaters and dry air in winter do not exactly improve the growing conditions for mushrooms, you will typically need to intervene during this time to ensure your cultures' needs are being met.

A cheap and easy option would be a plastic box with a lid. You can either place your entire cut-open growing kit directly into the box as-is or remove the substrate from the bag and place it on two small pieces of wood in the box. The idea is to create a microclimate in the box with higher humidity than the rest of the room. Spray the inner walls and the substrate as needed with a spray bottle. As mushrooms need enough oxygen to fruit, do not close the container completely.

Another extremely practical possibility is an indoor greenhouse, one of the small ones used to start seeds. There are many different sizes available, so you can choose one that suits the quantities you intend to grow. Since they can also be used for their originally intended purpose of seed starting in the spring, the small investment may be worth it.

High-Tech Indoor Greenhouses

Higher-tech versions of indoor greenhouses can also be useful for cultivating mushrooms. These can either consist of several gadgets linked together or purchased complete. These kinds of greenhouses are essentially made up of a large container with removable lid, a heating coil or mat with a thermostat, and a ventilator, which also regulates humidity.

In order to keep the humidity high in your greenhouse, you can cover the floor of it with moistened perlite, then place your culture on top of it. Perlite is a pH-neutral soil amender that can store large amounts of water. Furthermore, it is an entirely natural product that slowly gives off its moisture into its surroundings.

Such high-tech greenhouses allow you to more or less dial up the desired temperature and humidity. This shortens the colonization phase and makes for earlier harvests.

OUR TIP: *This investment is especially valuable for those who wish to grow medicinal mushrooms to replace mushroom powders and teas they would otherwise purchase.*

A Mushroom House for the Garden

A mushroom house for summer months in the garden can be just as useful as an indoor greenhouse in winter. They are especially good for those mushrooms that need extra heat and humidity for their cultivation. A mushroom

Indoor greenhouse

"PROTECTED" ENVIRONMENTS FOR ALL SEASONS

Mushroom house with reishi cultures

house is essentially a frame of some sort that is sheathed in plastic sheeting or double-walled polycarbonate sheets and can be erected in a shady spot outdoors. It should close up tightly and be easy to clean. A DIY mushroom house is great for growing mushrooms in pots, as well as for cultures growing on straw. Build it in a way that allows you to make the best possible use of every square inch (see photo above).

OUR TIP: *We have had excellent results bringing potted reishi logs into a mushroom house for fruiting. Even summer temperatures are too low for reishi where we live in Waldviertel, Austria. If this is also the case where you live, a mushroom house will be just the thing.*

Special Technique: Reishi in Pots

When growing reishi on hardwood (preferably beechwood), inoculate logs with the auger method. Just like for other cultures, logs are left in their full length while the culture colonizes them and, the next spring, the logs are cut in thirds. These cut logs can then be placed in pots (see also Setting Up a Mushroom Garden on the Balcony or Patio, p. 80), with cut faces covered with moss. Optimal conditions for reishi mushrooms prevail in a mushroom house in summer. You can also bury logs in the ground from spring until summer, then pot them up in midsummer. Mushrooms in the mushroom house will need to be watered, as they are being protected from rain.

Once the cap has developed well and the white growth zone has all but disappeared, you

Golden oyster mushrooms emerge from the growing medium

Lion's mane fruit body from a growing kit

Pink oyster mushrooms require higher temperatures

can cut away mushrooms. Reishi is notable for its ability to form new fruit bodies (see photos opposite). A new mushroom emerges from the mycelium remaining in the log after the first harvest, yet the mushrooms from the second flush are just as impressive as those from the first. In winter, store logs outdoors and bring them into the mushroom house again in the following summer.

"PROTECTED" ENVIRONMENTS FOR ALL SEASONS

Reishi in pots

Reishi regrowing after the first harvest

6. Pests and Competing Organisms in Mushroom Cultivation

Mushrooms are unfortunately not spared from pests and other competing organisms. The brief overview that follows indicates which organisms (in the garden, house, basement, cellar, or lab) can bring difficulties. Some of these are not necessarily cause for concern, but rather signs of a functioning ecosystem. The examples mentioned here are most relevant for hobby mushroom growers.

When speaking of pests, one usually means insects, mollusks, or mammals—and sure enough, from microscopically small mites to ravenous slugs, there are many undesired organisms that can affect mushrooms. Organisms like mold fungi, other wild competing fungi, bacteria, and even viruses can cause reductions in yield or even crop failure. The earlier you are able to recognize an infestation, the sooner you will be able to take countermeasures.

Mice

Mice are often found in the garden. They eat grain spawn out of the inoculation sites of log cultures (shiitake, buried logs). Freshly inoculated logs that are still being colonized (especially in winter) are common victims of mice. A "follow-up inoculation" doesn't make much sense. Once the colonization phase has ended (after about one year), grain spawn consumption by pests no longer does damage, as the mycelium is fully established throughout the log.

Countermeasures:

- Cover cut faces with aluminum plates (available from printing works), which seals in the grain spawn from mice. The aluminum can be stapled over the tape.
- Alternative: use auger method
- Appropriate siting for the stack of logs

In closed spaces, mice cause little damage. If you've installed a mushroom lab in your basement or cellar and use it for making spawn, store your grain in sealed containers.

Birds

Birds are sometimes tempted to pick old spawn from cut faces. This does not typically cause problems, as mycelium will have colonized the log by this time.

Slugs

Buried log mushroom cultures simply must be protected from slugs.

Summer mushrooms (Italian oyster, golden oyster, lion's mane, and reishi) are to be kept under close watch, because slugs tend to appear in greater numbers at this time. Mushrooms that fruit in spring or autumn are mostly safe

A ravenous slug devours a golden oyster mushroom

from slugs. For shiitake logs, hanging or vertically leaning logs can minimize slug problems.

Slugs love to just move right into mushroom cultures on straw bales and in beds. They leave eggs under straw in autumn. Summer mushrooms like king stropharia or various oyster mushrooms are common victims.

Countermeasures:

- For log cultures, slip water-permeable membrane or floating row covers over each log, held tight at the bottom with a strong rubber band and closed at the top (e.g., with a clothes pin). Allow plenty of space for mushrooms to grow unhindered (see photos, left). After harvesting, this covering can be removed.

OUR TIP: *Cut a rectangle out of the membrane and use adhesive to make into a "hose."*

- Another possibility is to surround the culture with slug-proof fencing
- If using a cold frame for your culture, cover with water-permeable membrane or floating row cover. Perhaps not slug-proof, but should significantly reduce numbers
- Slugs are active at night so hunt then

Fungus Gnats (of the Family Mycetophilidae or the Family Sciaridae)

Insects tend to become problematic in the garden when overripe fruit bodies stay on their growing medium for too long without being

Protecting buried cultures from slugs

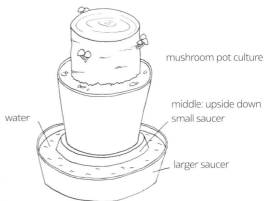

Protecting mushroom pots from slugs

harvested. It can be difficult to discern which family a given fungus gnat belongs to, but there are several countermeasures that are effective against both kinds.

Blocks of substrate in a mushroom house are vulnerable to infestation by fungus gnats. Slower-growing mushrooms are even more susceptible.

Countermeasures:

- Yellow panels can be highly effective. Affix some near your cultures to draw fungus gnats away.

There are many different species of fungus gnats in the Sciaridae family. Some only feed on mycelium, some on mushrooms. If they lay eggs, their larvae will likely turn up in the growing medium. Often times there is no choice but to remove the growing medium from the mushroom house.

In closed spaces (flat, house, basement, cellar), these insects may find their way to growing kits. The older the growing medium, the more vulnerable it is. Because fungus gnats prefer warm, moist conditions, they may also reside in the soil of potted plants. As they are not particular about soil, these gnats will also lay eggs in mushroom growing medium.

Infestations tend to be more intense in the summer.

Countermeasures:

- Yellow panels are also useful in closed spaces, as they draw fungus gnats away from your culture and help reduce damage
- As growing kits are especially appropriate for winter growing, you can use this timing to your advantage
- Overwetting your growing medium can increase your culture's vulnerability to infestation

- Cut open harvested mushrooms and check for larvae. Finding a few fungus gnats does not necessarily mean that the culture has been completely infested.

Mites

If you find mites in your straw, this is either reflective of the quality of the straw or the bale was overwatered, which encourages mite infestations.

Countermeasures:

- Use only healthy straw with minimal weed content
- Protect from the rain to avoid excessive moisture
- Complete, thorough fermentation of the straw when growing oyster mushrooms, which discourages mites

Microscopic view of a mite crawling in a net of mycelium

Once mites have been brought into a mushroom cultivation lab, you will need to be obsessive about hygiene and all affected cultures will need to be removed. Mites eat mycelium and are difficult to find because they are so small.

Competing Wild Fungi

It is not uncommon to find other fungi on logs such as split gill, birch maze-gill, shelf fungi, polypore species, or woodwarts, all of which appear only temporarily. They are usually easily differentiated from your cultivated mushroom; there is little danger of confusing them. These fungi colonize parts of the wood that the cultivated mushroom does not and contribute to the natural decompositional process of the log.

Birches are especially vulnerable to competing fungi. Older beechwood logs host so-called woodwarts (*Hypoxylon*) and slime molds (wolf's-milk fungi). These spherical, approximately 3/16 in (5 mm) mushrooms are only found on the bark and are no cause for alarm. They can easily be removed with a sharp knife at the right moment in their development. There are several different species that may appear.

Covering buried logs with moss does have the disadvantage of introducing spores from species other than the one you are cultivating. But, since the protection they provide against drying out is so valuable, we gladly tolerate the presence of the occasional harmless secondary species.

Where hardwood chips are being used as ground covering, competing wild fungi can establish themselves. Old logs are easily colonized by nonedible fungi that were previously established in the area.

If logs have been infected with sulphur tuft or honey fungus, remove them immediately from your mushroom garden to prevent their spreading.

Woodwarts (Hypoxylon)

Wolf's-milk fungus

94 HOME-GROWN MUSHROOMS FROM SCRATCH

The diversity of secondary fungi is vast

OUR TIP: *Always confirm that it is in fact your cultivated mushroom that you are harvesting and not a wild fungus species that has outcompeted yours. This is another reason why labeling your cultures is important. Pay special attention to your first harvest. Once you have become deeply familiar with the mushrooms you are growing, you are less likely to mistake another variety for it.*

Inky Cap, Cup Fungus, Yellow Fieldcap Mushrooms

Straw-based cultures may be affected by secondary fungi like inky cap, various cup fungi, and yellow fieldcap mushrooms. If, despite the mycelium of your oyster mushroom or king stropharia having well-colonized the straw, other mushrooms appear temporarily, remove these to prevent their spores from spreading. These secondary species are not edible.

When growing on straw pellets, cultivated mushrooms typically have no problem with competing fungi. Should inky cap nevertheless appear, remove the culture.

Mold

Overwatering mushroom cultures in the mushroom house can cause mold to spread. Simply having high humidity in an enclosed space can also be enough to allow mold to thrive.

Countermeasures:

- Apply salt to mold or use vinegar to remove
- Increase air circulation and access to fresh air
- Thoroughly clean mushroom house before starting new cultures—wash all surfaces and disinfect with 70 percent isopropyl alcohol to keep mold at bay

Countermeasures:

- If the entire log is not affected, fruit bodies can simply be left on
- Remove any logs where your mushroom culture has not been able to establish itself at all and only secondary fungi grow

PESTS AND COMPETING ORGANISMS IN MUSHROOM CULTIVATION   95

Molds are rare on wood-based growing media. Healthy straw and balancing humidity levels are generally sufficient to keep mold away.

Countermeasures:

- Cover cultures in case of rain
- Do not overwater
- Harvest thoroughly so that no mushroom pieces are left behind, as these are potential points of entry for mold and pests

Mold can, unfortunately, also appear in the lab. When working under semi-sterile conditions, the risk of contamination is higher.

Be particularly careful with fertilized substrates (with substances like bran, sugar, etc.). The more sugar and nitrogen in a growing medium, the more vulnerable it is to being contaminated by mold or bacteria.

Countermeasures:

- Work hygienically
- Sterilize growing media in pressure cooker or autoclave
- Keep working surfaces clean

OUR TIP: *Find more on this topic in the chapter entitled "Propagating Mushrooms—From Spore to Spawn," p. 96.*

Growing kits in enclosed spaces are also vulnerable to mold. Overwatering and a site already infected with mold are common reasons for moldy growing kits.

Countermeasures:

- Remove visible contamination: apply salt or vinegar, scratch moldy patches, or remove growing medium
- Do not allow substrate to get too old

Cup fungi

Inky caps in various stages of development

Mold fungus on an agar medium

Lion's mane under the microscope

7. Propagating Mushrooms— From Spore to Spawn

For Specialists, Experienced Growers, and Novice Experimenters

Up until now we have focused quite a bit on the various materials of mushroom cultivation (grain spawn, plug spawn, growing kits, etc.). In order for you to understand the life cycle of fungi, we shall now take a closer look at the individual development stages that a cultivated mushroom species goes through in the course of its life. There are several prerequisites for successful cultivation:

- a clean, germ-free workspace
- knowledge of microbiological techniques
- appropriate tools
- laboratory devices
- the ability to work in a concentrated, systematic way

An Overview of the Stages of Mushroom Propagation

- Setting up your own mushroom lab
- Internalizing hygienic practices and work stages
- Producing growing media
- Extracting pure culture (via tissue sample)
- Producing grain spawn
- Producing plug spawn

Each of these work stages requires other materials and knowledge. And the costs of inventory and working materials are not to be forgotten. Resources are listed in the appendix. A general overview can be found in the table "Working Steps and Materials" on p. 100. Before we go into more detail, we would like to expand upon a few terms.

What does it mean to be hygienic in the workspace? When we speak of a hygienic workspace, there are several important factors. The area must, of course, be kept clean. Ideally you will choose a space to use as a mushroom cultivation lab that is easy to clean and otherwise empty. It is worth it to put some time and effort into the design and preparation of a cultivation room. The air is full of myriad microorganisms—if the airborne-germ pressure becomes too high or if there is a lot of movement in the room, it will be difficult to maintain the required standard of hygiene. Rugs, curtains, and other dust catchers must be removed and draughts must be avoided. Tiled rooms make for excellent mushroom labs as they are easy to clean and you can use a disinfectant on them. We recommend working surfaces of glass or stainless steel. People are potentially important contamination vectors. Wear the cleanest clothes possible and use hand sanitizer.

Why disinfect? Fungi are grown on media that could also potentially host many bacteria, yeasts, and other organisms. The idea is to eliminate all possible sources of contamination in order to give your cultivated mushrooms the best chance to thrive.

How does one make working surfaces, materials, and tools germ free? A wide variety of disinfectants are available commercially (see Sources, p. 137). One of the simplest disinfectants is 70 percent isopropyl alcohol, which you can fill a spray bottle with. Simply spray surfaces, use a paper towel to spread evenly, and then just wait a moment for it to evaporate. (There should be no alcohol left over, as it is flammable.) Wash your hands afterward and finish up with hand sanitizer. You can also use disposable gloves (the outsides of which should be disinfected).

Tools and containers (scalpel, inoculation loop, petri dishes, etc.) can be handled in different ways. Before and after each step, tools must be singed with an alcohol lamp or blowlamp. Singe not only the tip, but a good two-thirds of a scalpel (warning: do not burn your fingers!). Do the same with an inoculation loop.

Sterilize glass petri dishes in a pressure cooker or autoclave. Otherwise, buy sterile plastic petri dishes. There are also sterile single-use utensils available. If germ-free growing media (grain spawn, growing kits, nutrient agar) are to be made, a pressure cooker and/or an autoclave are a must.

What are pressure cookers and autoclaves used for? In a closed container, water can be heated to above its boiling point, which creates excess pressure. In a pressure cooker, the pressure is 0.8 bar and the temperature is 241° F (116° C). In an autoclave, pressure and temperature are higher (1 bar, 250° F [121° C]).

These conditions ensure sterilization. The timing of sterilization will depend on circumstances.

What tools and containers are used in lab work?
- scalpel
- inoculation loop
- pincers
- spray bottle with alcohol
- indelible marker (for labeling)
- petri dishes (glass or disposable)
- alcohol lamp or blowlamp
- autoclave bags of various sizes
- jars with filter discs
- Erlenmeyer flasks

What is a glovebox and what is a Laminar Air Flow? The cheapest option for microbiological work is to build your own glovebox, which will

Tools needed in any mushroom lab

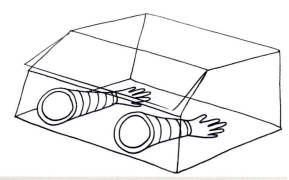
Glovebox

allow you to work in semi-sterile conditions. To be able to work effectively with a glovebox, it should be customized to the person using it and the space it is being used in. A glovebox creates a small enclosed space with minimal air circulation and can easily be disinfected. Important: No flame should ever be lit in a glovebox!

A basic DIY glovebox is essentially a sealable, transparent plastic box. Cut two holes in the box and affix long gloves to them to keep out germs.

A Laminar Air Flow with a HEPA (High Efficiency Particulate Air) filter eliminates up to 99.99 percent of airborne germs. The filter, in combination with a blower, produces a "sterile" airflow. You can then work without complications in this milieu. This is quite an expensive piece of equipment, however. Many laboratories have HEPA filters in use. If you have the time, you can just buy the filter and install it in a box with a blower.

You should ponder how much you would like to invest in a mushroom lab before sinking a lot of money into it. For those that have only recently begun to cultivate mushrooms, a glovebox may make for a simple, cheap, and practical acquisition.

What growing media are used in mushroom cultivation? Agar, a jelly-like substance derived from algae, is the first medium used in cultivating. It is used to grow mycelium from mushroom tissue or spores. Malt extract agar and potato dextrose agar are the most commonly used. There are different recipes for these media. Antibiotic nutrients are now also available (not always recommended, as the antibiotics used can affect the growth of some kinds of mushrooms). The pH value of the growing medium is also relevant. Most mycelia are fairly tolerant and can grow in pH values ranging

PROPAGATING MUSHROOMS—FROM SPORE TO SPAWN   99

Growing medium ingredients, from the top: rye, millet, beechwood furniture dowels, beechwood shavings, straw pellets

from 5.5 to 7.5. Later, other growing media are used such as various grains, straw, hardwood chips, and hardwood furniture dowels.

Grain spawn is a mixture of several ingredients, for example: organic rye, organic millet, hardwood shavings, and additives like gypsum. Hardwood furniture dowels are used to produce plug spawn. Every producer has their own special mixture.

Is it possible to skip certain steps or does one always have to start at the beginning? Several components can now be purchased from specialist stores. Everything from sterile disposable utensils and disposable petri dishes to sterile nutrient agar and other growing media are available commercially. For growers, this can make work easier on the one hand, while on the other hand make growing more expensive.

Does one need an advanced degree in microbiology to operate a mushroom lab? Growing and cultivating mushrooms requires a broad range of skills. There are no limits to the degree to which you can involve yourself. Start small and slowly become more advanced in your undertakings. In this way, your experiences are more likely to be successful and you won't lose motivation because of being in over your head.

Practice makes perfect—gaining experience with the various stages of cultivation, thinking about them, and internalizing them are all extremely valuable. The more you proceed in this way, the fewer mistakes you will make and the more advanced your cultivation techniques can become.

Working Steps and Materials

For Novice Experimenters

Task	• make plug spawn by simple means (p. 105)
Required materials	• grain spawn • pressure cooker • glass jars or autoclave bag • hardwood furniture dowels • malt extract • gauze (filter) • aluminum foil
Hygienic demands	• relatively low • clean, dust-free, disinfected workspace

For Experienced Growers

Task	• set up a lab and do microbiological work • make nutrient agar
Required materials	• sterile environment: glovebox or Laminar Air Flow (HEPA) • 70 percent isopropyl alcohol to disinfect surfaces and hands • sterile glass petri dishes or disposable petri dishes • Erlenmeyer flasks • scale • measuring cup or beaker • pressure cooker • malt extract agar (mix)
Hygienic demands	• high • glovebox or Laminar Air Flow must be acquired (consider costs)

For Specialists

Task	• extract a pure culture • make grain spawn • make growing kit
Required materials	Same as for Experienced Growers, plus: • scalpel and inoculation loop • alcohol lamp or blowlamp • growing medium ingredients: rye, millet, gypsum, hardwood shavings • heat sealer • pressure cooker or autoclave • glass jars with filter discs or autoclave bag
Hygienic demands	• very high • Laminar Air Flow (consider costs)

An Overview of the Working Stages of Mushroom Propagation in the Lab

- make nutrient agar
- select a healthy mushroom
- make a tissue culture: tissue sample on malt extract agar
- growth phase: mycelium spreads
- make grain spawn: transfer mycelium to a sterile growing medium
- make plug spawn

Pouring liquid agar into a petri dish

Making Nutrient Agar

As previously mentioned, there are many different recipes for nutrient agar. The mix we give here can be used for most cultivated edible mushroom species and will yield enough for about ten petri dishes. Malt extract agar is available as a powder. Mix it with water in an Erlenmeyer flask in the following ratio:

- 0.44 oz (12½ g) malt extract agar to 1 cup (¼ L) water
- Important: prioritize the agar producer's instructions over ours (in terms of proportions)

OUR TIP: *Start by adding the agar to the flask, then add water quickly and swish about to avoid clumps.*

Close the opening of the flask with aluminum foil and then place in boiling water for 15 to 20 minutes. This allows the agar to completely dissolve in the water. Thereafter the liquid must be sterilized, using either a pressure cooker or an autoclave. If using an autoclave, follow the instructions included with it. For a pressure cooker, place the flask inside and then fill the cooker ¾ in to just over an inch (2 to 3 cm) deep with water. Close and heat. When the right temperature has been reached and steam vents out of the safety valve, cook for 20 more minutes to sterilize.

Finally, remove the cooker from heat and allow to cool slowly. Do not let it cool completely, as the agar will stiffen. You should be able to touch the flask once it cools to about 104° F (40° C), yet the agar will still be liquid.

Pouring Agar into Petri Dishes

Before beginning:

- work in a glovebox or in front of a Laminar Air Flow
- wash and sanitize hands
- clean workspaces: spray 70 percent isopropyl alcohol, spread with paper towel, let dry
- ready sterile disposable or sterile glass petri dishes at your workspace

Remove aluminum foil from the flask. Remove petri dish lid and quickly pour in malt extract agar. Fill the first dish about $3/16$ in (5 mm) deep or $1^{2}/_{3}$ tbsp (25 ml), then replace the lid and proceed with the rest of your petri dishes in the same fashion.

OUR TIP: *The liquid should distribute itself across the entire floor of the dish. When you have filled them all, the petri dishes with agar should cool (ideally in front of a running Laminar Air Flow). Condensation may form on the inside walls of the petri dish, but this should disappear by the time you move on to the next stage.*

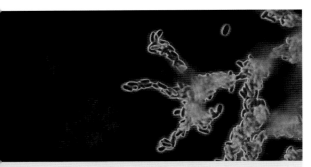

Pleurotus pulmonarius spores

Selecting a Healthy Mushroom or Spore Print for Making a Pure Culture

If you would like to propagate a fungus, this can either be done from a fresh tissue sample or from a spore print. Spore prints can be stored for a relatively long time. The danger, however, is that it may be impure. As a countermeasure, you can use an antibacterial growing medium when propagating mycelium. This method of propagation is also used when figuring out the genetic diversity of a given fungus. Individual strains crystallize out of the mycelial network, which can then be selected and further propagated. This is comparable to the selection process when saving seed for plants; you select seeds from those plants with desirable traits like flavor, disease resistance, or high yield.

If you are working with a tissue sample, there are no variations (the mycelium is all genetically identical). Once you have picked a mushroom for propagation, there is no more selecting to be done.

Before you get to work:
- work is to be done in front of a Laminar Air Flow
- wash and sanitize hands
- clean workspaces: spray 70 percent isopropyl alcohol, spread with paper towel, let dry
- disinfect utensils (scalpel) before each use or use packaged sterile blades
- have cooled agar petri dishes at the ready
- select healthy mushroom (if not using mushroom immediately, store cool and clean for no more than two days)

Making a Tissue Culture

Cut the stem of the mushroom with the scalpel about $3/16$ in (about 0.5 cm) deep and then carefully pull the mushroom apart along its length. Your fingers should never touch the innards of the mushroom. Now remove a piece from the inside of the mushroom, as, unlike on the outside, other organisms will not be found here. Cut a piece up to $3/16$ in (about 3 to 5 mm) from the area near the base of the cap with the scalpel. Carefully spear the piece with the scalpel and as quickly as possible, in one motion, deposit the tissue sample in the middle of an agar petri dish.

When transferring the tissue sample onto the nutrient agar, lift the lid of the petri dish for the briefest amount of time possible. After you have deposited tissue samples into all your petri dishes, label each on the lid with the name of the mushroom and the date to eliminate the possibility of confusion later on.

OUR TIP: *To keep the risk of contamination at an absolute minimum, avoid bringing your hands between the air filter and your work. Otherwise germs could more easily land on the nutrient agar.*

Growth Phase

Now the mycelium can grow on the malt extract agar. Finding an appropriate place for the petri dishes is important. They should be moved as little as possible. In a closed case in front of the Laminar Air Flow would be the ideal location. Alternatively, a disinfected plastic box should also work well. Within three to

PROPAGATING MUSHROOMS—FROM SPORE TO SPAWN **103**

Halve the mushroom . . .

. . . cut out a piece of tissue . . .

. . . remove it from inside the mushroom . . .

. . . and deposit on sterile nutrient agar in a petri dish.

four weeks, mycelial growth should be clearly visible and it should ultimately reach out to the edge of the dish.

If mold should appear on the nutrient agar, remove it immediately. If the cultures have been affected, do not open the lid—this could allow the spores to spread. Losses of 10 to 20 percent are considered tolerable in the world of hobby mushroom propagation. Theoretically, several pieces from one successful petri dish culture can be used for further propagation.

Each kind of mushroom develops mycelium in its own idiosyncratic way. The distinctive smell of each mushroom is also recognizable at this point. Growth and appearance are dependent to a degree on environmental factors. Store the cultured petri dishes in darkness.

Agar petri dishes are not appropriate for longer-term storage—a special long-term culture will need to be established. Pour agar into test tubes held askew, then transfer mycelium onto it. When these tilted agar cultures are stored in the refrigerator, they will keep for several months.

Making Grain Spawn

The next step in the propagation process is the transference of mycelium from a petri dish onto a grain substrate. This is how grain spawn—the base material for inoculating the final, mushroom-producing substrate (wood, straw, mushroom bed, compost)—is made.

Recommended substrate recipe:
- 1 part organic rye
- 1 part organic millet
- 1 part fine beechwood shavings
- 0.176 oz (5 g) gypsum per 2.2 lb (1 kg) spawn (add after cooking)

Substrate recipes vary from grower to grower. Mix all the above ingredients except the gypsum with sufficient water and cook. It is important

Grain spawn in various phases of growth

Gauze can be used as a filter

Lid with filter disc

to regularly check the consistency of the grains (they should not become too wet, nor should they burst). Water should lightly boil as you cook it, stirring occasionally. Once the grains of rye are soft but the bran is still sturdy, strain out the excess water. Now is the time to mix in the gypsum and then fill jars up to ⅔ full with the mixture. Cut a circle out of several layers of gauze for a jar lid (see photo, top right). Attach the gauze to the jars with a $3/16$ in (5 mm) thick silicone hose, which must be heat resistant. Alternatively, you can make a hole in a jar lid and glue a filter disc to its inside (see photo, bottom right). Wrap these two-thirds-full filter jars with aluminum foil. This keeps moisture out of the jars while they are in the autoclave and further reduces risk of contamination.

Mushroom grower supply dealers have autoclave bags with an integrated filter. These can be used instead of jars. They are available in sizes ranging from approx. 1 to 4 quarts (1 to 4 liters). Once you've filled a bag, fold it such that air gets purged. Then sterilize in a pressure cooker or autoclave for about two hours—the pressure cooker should have 2 to 3 cm of water at the bottom so that steam is present during sterilization. Afterward allow the substrate to cool.
Before getting to work:
- work in front of a Laminar Air Flow
- wash and sanitize hands
- clean workspaces: spray 70 percent isopropyl alcohol, spread with paper towel, let dry
- disinfect utensils (scalpel) before each use or use packaged sterile blades
- have petri dishes with propagated mycelium at the ready

With a scalpel, cut up the agar into six pieces of equal size.

Remove a cooled jar or bag of sterilized grain substrate from the pressure cooker and carefully open—do not let your fingers touch the rim (contamination risk). Carefully spear a piece of agar and transfer into the jar with the mycelium side oriented inward. Replace gauze or filter lid.

Transferring multiple pieces into a single glass will speed colonization of the grain substrate. If you are using autoclave bags, you will need a heat sealer to properly seal them.

PROPAGATING MUSHROOMS—FROM SPORE TO SPAWN

Cut several pieces out of the agar

Spear onto scalpel

Transfer to sterile growing medium

Transferring a piece of agar

Colonization Phase

It is ideal to place the agar pieces in the middle of the substrate (deep inside). Soon, fine, white strands of mycelium will be visible if colonization is proceeding successfully. From time to time, you may shake the culture to more evenly distribute the mycelium.

After four to five weeks—depending on mushroom type, temperature, and humidity—the mycelial network should be well-established and the substrate will have been fully colonized. If you find mold at this point, the culture will unfortunately have to be disposed of.

Making Plug Spawn by Simple Means

Further propagation via grain spawn is something for eager experimenters. The individual steps can be executed without much effort and without expensive accessories. The necessary tools can be found at the hardware store if you do not already have them. The only thing you will have to order will be grain spawn, which will be used for inoculating and also for producing plug spawn.

Before you get to work you will need:
- to order desired grain spawn (shiitake, oyster, king stropharia, etc.)
- pressure cooker
- heat-tolerant jars (with an opening of about 2¾ in [7 cm])
- untreated hardwood furniture dowels (ideally beechwood; found at hardware stores)
- malt extract (found at pharmacies)
- gauze as filter for jars
- aluminum foil
- spoon
- string
- rubber band

Just 1 pint (½ liter) grain spawn will be sufficient to inoculate 3.3 lb (1.5 kg) worth of hardwood

Sterilized furniture dowels next to grain spawn

Transferring grain spawn to furniture dowels

furniture dowels. Make sure the dowels are of untreated wood (not treated with fungicides).

Place dowels in a pot with sufficient water (they should soak up as much water as possible) and bring to a boil. You may enrich by adding 0.35 oz (10 g) of malt extract per quart (liter) of water. Make sure everything you use in this process is clean to prevent bacteria and yeasts from moving in.

After about 30 minutes of boiling, allow the pot to cool and pour off any excess water. Fill jars about two-thirds full with the dowels. Cut circles out of several layers of gauze such that they overlap the rim and tie on with a string. Finally, cover the jars with aluminium foil. Sterilize the jars and a foil-covered spoon in a pressure cooker. Fill the cooker 2 to 3 cm deep with water, sterilize for 20 minutes, then let cool.

Clean a space to work on and transfer grain spawn into sterilized jars. Cover with gauze and affix with a rubber band.

Store the jars in a dark, clean place. Colonization will take about 4 weeks. Mycelium spreads from the grains onto the dowels (for details about inoculating logs or stumps with plug spawn see "Mushrooms from Logs," p. 18).

Ganoderma lucidum *on beechwood*

Reishi in various stages of processing

8. The Use of Mushrooms in Medicine

There are a plethora of mushrooms whose metabolic products can be used to fight viruses, bacteria, and harmful fungi. Probably the most well-known example is penicillin, which is the metabolic product of a mold fungus. Another is lentinan, isolated from shiitake mushrooms, which is used in cancer therapy.

Traditional Chinese medicine (TCM) has employed various medicinal mushrooms for millennia in the prevention and treatment of disease. Among the many advantages of medicinal mushrooms are that they are nearly free of side-effects, can be used preventively, and can restore homeostasis in the body. Europe also has a tradition of medicinal mushroom use. We also know that native mushrooms were long used to maintain and restore health in monastic medicine. This knowledge has unfortunately been pushed into the background but is slowly becoming appreciated again. Recent decades have seen intensive pharmaceutical research into mushroom compounds. Several interesting articles on this topic can be found on an internet portal for scientific studies called Pubmed (see Bibliography, p. 138).

Mycotherapy is an important field involving medicinal mushrooms and falls under the category of naturopathy. It is used to supplement other holistic forms of therapy. If you are interested in this alternative medicine technique, we encourage you to get in touch with a mycotherapist or a TCM practitioner. Before consumption and use, your tolerance for a given mushroom should be evaluated. Most medicinal mushrooms are well-tolerated, however, and can be used in diverse preventative therapies or in supporting good health.

The mushrooms that we describe can be grown by yourself, found wild, ordered from specialist retailers, or found at a health shop. Some of these mushrooms have very complex life cycles or are very difficult to cultivate.

There are several institutions that offer a wide selection of medicinal mushrooms and offer competent answers to questions about their use.

OUR TIP: *Bibliographical entries are listed in the appendix for further reading. There you can also find sources for medicinal mushrooms.*

Reishi Mushrooms, Lingzhi Mushrooms (*Ganoderma lucidum*)

Reishi is one of the most highly prized medicinal substances in TCM. The Chinese called it the "godly mushroom of immortality" or "plant with spiritual powers." Its name indicates its preciousness and its medicinal potential. Lingzhi is native to Japan, China, and Europe. Its growth habit is reminiscent of polypores, and its cap is cork-like and firm. Depending upon where it is in development, it can have quite a quirky appearance. Reishi mushrooms can be brownish-red to orange-white-yellowish in color. They have a bright yellowish-white growth zone that is quite sensitive, and you can use this to follow its growth. When reishi mushrooms are dried, they keep their shape and hardly shrink at all. Reishi is a wood-loving, saprophytic fungus that colonizes deciduous trees and cultivates fairly well on beechwood logs (see Special Technique: Reishi in Pots, p. 87). Alternatively you can use a growing kit or simply order the mushroom.

Because of its woody fruit body and its slightly bitter taste, it is most typically used as tea, in extract form, or pulverized in capsules.

Its triterpenes and beta-glucans are the most prized metabolic products of this mushroom. They are immune stimulating, used therapeutically in cancer treatment, and can prevent various illnesses. Studies have shown its positive effects on cardiovascular disease as well.

An overview of the fields of use for reishi:
- cardiovascular disease
- blood pressure regulation
- adapting to altitude
- modulating and strengthening the immune system
- adjunct cancer therapy (stomach, liver, lung, skin, brain, kidney)
- allergies
- infections (particularly chronic hepatitis)
- inflammation
- asthma, bronchitis
- insomnia, fatigue
- anti-aging

OUR TIP: *When cultivating and processing reishi yourself, observe the following: Cut mushrooms when fresh and dry completely in a dehydrator or oven (max. temp. 104–122° F [40–50° C]). It is essential to then store in containers that seal completely. It can then be ground as needed for tea in a coffee mill, the finer the better, one to two teaspoons of reishi powder per cup of tea. As reishi tastes somewhat bitter, you can mix a smaller amount of it with other teas or mix it in with your coffee. Do not pour boiling water over reishi powder, as this could damage heat-sensitive compounds contained therein.*

When buying reishi powder or extracts, ask for products that come from organically grown mushrooms and that have been processed gently. Dosing should ideally be done on a person-by-person basis by a doctor or mycotherapist. Recommended daily doses are often found on the packaging, which can be helpful when being used to prevent illness.

Shiitake is easy to cultivate

Shiitake (*Lentinula edodes*)

Shiitake is known for its excellent flavor and the broad spectrum of therapeutic contexts in which it can be employed. It contains protein, fiber, vitamins, and minerals. It has a characteristic aroma that is also present in dried mushrooms. There is currently much research being done into the polysaccharide lentinan and the chemical compound eritadenine, both of which are contained in the shiitake fruit body.

The polysaccharide stimulates the immune system and contributes to the production of special antibodies. It has been used successfully with HIV-infected patients. Lentinan is being used in Japan to support the treatment of tumors. This compound is also an ingredient in a cancer medication used in the United States. Lentinan has also shown itself to be useful against viral infections (e.g., influenza). Eritadenine helps to regulate blood lipid levels. Consuming shiitake can have positive effects on your cholesterol levels. Shiitake can also enhance your nutrition and psychological well-being, as it contains vitamin D and its provitamins.

An overview of the fields of use for shiitake:
- immunodeficiency
- common cold, flu
- infectious diseases (bacterial and viral)
- regulation of fat metabolism
- osteoporosis
- gout
- arthritis, fibromyalgia
- concomitant with cancer therapy
- supporting intestinal microflora

OUR TIP: *On page 118 you will find instructions for how to preserve shiitake mushrooms. Important information: Sensitive people may develop skin rashes when consuming mushrooms. This will appear very soon after eating; fortunately there are very few cases of shiitake-dermatitis. In the case of an inflammatory reaction, use another medicinal mushroom. Skin blemishes may also appear temporarily—these are signs of the positive processes of drainage and detoxification.*

Lion's Mane/Bearded Tooth (*Hericium erinaceus*)

Another medicinal mushroom, one that is not known as much for its aroma as its positive effects on the body, is known commonly as

THE USE OF MUSHROOMS IN MEDICINE

lion's mane or bearded tooth mushroom. Its icicle-like growth and its creamy-white to pink coloration make it into a real sensation in the garden. It can be grown on hardwood (beech, birch, oak) or from a growing kit. It is a true delicacy of Asian cuisine.

Indigenous people of North America used this mushroom as a hemostatic for cuts. It also has positive effects on gastrointestinal diseases and nervous diseases.

The compounds contained in the fruit bodies of *Hericium* have been fairly thoroughly studied. The mushroom contains all eight essential amino acids. Further, it contains much potassium and other important minerals like zinc, iron, selenium, and germanium. The polysaccharides and polypeptides it contains have also caught the attention of researchers of *Hericium*. They have been shown to support the immune system and increase the body's own defenses against antigens. The mushroom has a stimulating effect on the regeneration of nerve tissue. Extracts of the mushroom are used for this reason in the treatment of Alzheimer's disease and Parkinson's disease. A brief overview of the fields of use for lion's mane:

- gastrointestinal problems (inflammatory, functional)
- cancer prevention, especially for stomach, intestinal, esophageal, and skin cancers
- nervous disorders
- anxiety, depression
- inner unrest, insomnia
- immunomodulation
- chronic skin conditions

OUR TIP: *Lion's mane is an excellent and very special edible mushroom—don't pass up the chance to cultivate on a log or grow from a growing kit. Recipes can be found on page 118.*

Lion's mane being cultivated on a growing medium

Shaggy Mane (*Coprinus comatus*)

This mushroom belongs to a small genus of fungi whose gills autodigest, releasing a black, spore-filled liquid ("ink") (see photos, p. 112). This only happens if you harvest too late, however, so be aware of this when cultivating yourself.

Coprinus comatus mushrooms have balanced vitamin and mineral content (with especially high vitamin C levels). This makes it interesting as a medicinal mushroom. Thanks to its ability to lower blood sugar levels, it has been used to help treat diabetics. Shaggy mane has been fairly well researched in Europe, where it is native. Important nutrients contained therein include potassium, magnesium, iron, calcium, manganese, zinc, copper, and vanadium. The latter is also involved in reducing blood sugar levels. In TCM, shaggy mane is used to prevent hemorrhoids and to encourage digestion. It also regulates the metabolism.

Shaggy mane mushrooms can be found in meadows . . .

. . . however, they quickly turn black, secreting spore-filled "ink"

An overview of the fields of use for shaggy mane:
- diabetes (reduces blood sugar in type I and especially type II)
- arteriosclerosis, circulation irregularities
- digestive problems, problems with intestinal flora
- hemorrhoids
- sarcoma (swellings on connective and sustentacular tissue)
- inhibition of breast cancer cells (regardless of estrogen receptor type)

OUR TIP: *In the chapter "Speciality: Shaggy Mane in Beds and Containers" (p. 72), you will find examples of methods for shaggy mane cultivation.*

Almond Mushrooms (*Agaricus blazei Murrill* [ABM])

This mushroom is a relatively recent discovery, but its incredible effect on human health was quickly recognized. Its origins are also special—the almond mushroom comes from the rainforests of Brazil. Farmers in this region first gathered these mushrooms from the wild, then eventually started to grow them. Researchers became aware of almond mushroom and were able to tease out some of its positive attributes. Almond mushroom is also popular in Asia. *Agaricus blazei Murrill* belongs to the same family as the button mushroom and likewise prefers fermented growing media. It thrives on high temperatures, high humidity, and a bit of sunlight. For this reason, cultivation of this mushroom is most appropriate for more experienced growers.

The mushroom has proven preventative anti-carcinogenic properties and also has anti-allergic, anti-diabetic, and anti-inflammatory effects on the body. Several scientific studies have shown their influence on the degeneration of tumors. These studies were

THE USE OF MUSHROOMS IN MEDICINE

Almond mushrooms from a growing kit

A young wood ear mushroom grows

done with ABM extracts. Additionally, almond mushroom helps minimize the side effects of chemotherapy. Studies have confirmed almond mushroom's role in inhibiting metastasis in leukemia, intestinal, lung, throat, breast, pancreatic, liver, and prostate cancers.

An overview of the fields of use for almond mushrooms:
- cancer prevention
- immune deficiencies, immune imbalance
- skin diseases
- digestive problems, intestinal regulation
- diabetes
- high blood pressure, issues with fat metabolism
- allergic reactions
- minimizing the side effects of chemotherapy and radiation therapy

OUR TIP: *If you would like to cultivate almond mushroom yourself, growing kits in an indoor mini-greenhouse are an easy way to go. In this warm, moist environment, mushrooms will be ready to harvest fairly quickly. Information on how to handle an indoor mini-greenhouse can be found in the chapter entitled "'Protected' Environments for All Seasons" (p. 85).*

Wood Ear/Jelly Ear (*Auricularia polytricha*)

This mushroom is well-known in Asian cuisine and it is also native to large parts of Europe. The fruit body of wood ear often grown on weakened elder shrubs and on other deciduous wood such as willow. You can cultivate them on logs or from a growing kit. Otherwise you can collect them in the wild.

Its fruit bodies reach a size of $1/8$ to $3/16$ in (3 to 10 mm). They can be quite variable in color, from light brown to reddish-brown.

Thanks to its appearance, which resembles the external part (pinna) of a human ear, and its gelatinous texture, it is often easily identified in the wild. Still, you should always confirm its identity with a mushroom identification book.

The mushroom is used to improve blood flow and generally encourages good circulation without affecting blood vessels. Blood-thinning medicines often have a negative effect on blood vessels, which makes the use of *Auricularia polytricha* particularly valuable. Experiments with water-based and alcohol-based extracts have shown an anti-thrombotic effect.

Wood ear growing from a kit

An overview of the fields of use for wood ear:
- arteriosclerosis
- blood pressure regulation, high blood pressure
- reduction of blood clotting, anti-thrombotic therapy
- arterial blood flow disorders
- myocardial infarction (risk reduction, prevention)
- immune system modulation, immune regulation
- eye, skin, or mucous membrane infections

OUR TIP: *Wood ear is excellent for thickening soups and sauces and adding special flavor when used in its ground form. Auricularia polytricha can be found nearly year round on weakened trees—just go for a walk in the woods and keep your eyes open. Cultivation can be tricky; it might not be for you if you want nearly guaranteed results.*

Oyster Mushrooms (*Pleurotus ostreatus*)

Oyster mushrooms are not only easy to cultivate, but they also contain many compounds with positive effects on the human body. Native to Europe, they grow on various deciduous trees (beech, alder, willow, etc.). Like other mushrooms listed above, oyster mushrooms are rich in vitamins, especially those from the B complex. Vitamins C and D in addition to folic acid are among its important nutrients. Vitamin D supports bone development and plays a role in preventing osteoporosis. Folic acid is involved in hematosis, the growth of young cells, and the lowering of homocysteine values in the blood. All these components help reduce the risk of cardiovascular diseases. *Pleurotus ostreatus* has anti-viral properties and can contribute to the lowering of cholesterol and triglycerides in the blood and the liver. Experiments have shown support for the growth of probiotic bacteria in the intestine.

An overview of the fields of use for oyster mushrooms:
- relaxes muscles, tendons and joints
- lowers risk of osteoporosis
- supports intestinal flora

OUR TIP: *Oyster mushrooms are a great addition to a healthy diet. They are easy to cultivate and have diverse uses in the kitchen. For more on cultivation, see Mushrooms from Logs, p. 18.*

Birch Polypore (*Piptoporus betulinus*)

This is an inconspicuous mushroom, and you probably pass many of them when walking through the woods. It parasitizes weakened birch trees or colonizes recently deceased ones. Once a birch tree has become infected with it, limbs will break off more easily. If hunting wild birch polypore, you should familiarize yourself with the most important identifying features.

Depending on the stage of development, it can reach 4 to 12 in (10 to 30 cm) in size. The fruit bodies have a creamy light brown coloration on top of the cap and fine white pores below. The flesh is compact, white, and dense. It has a handle-like, convergent base, and as it grows, it looks like a pillow growing out of the side of a tree.

They appear throughout the year, so you'll have to figure out how fresh they are. Birch polypores should be harvested when white under the cap. As soon as dark spots and a dark underside are recognizable, it is no longer harvestable. The topside of the mushroom feels smooth. Fruit bodies are best gathered in summer through late autumn. Pores can be easily separated from the flesh when the mushroom has grown old. Sometimes, birch polypores have a lightly waved edge. When still young, they have a bulbous appearance (see photo, p. 116, top).

Birch polypore

According to mushroom consultant Hans-Heinrich Kunde, birch polypore is helpful in cases of weakness of the stomach, indigestion, stomach irritation, and gastritis. It can be drunk as tea for prophylaxis against the above maladies. Polyporenic acid, found in the fruit body, is anti-inflammatory and anti-bacterial.

An overview of the fields of use for birch polypore:
- indigestion
- bloating, gas
- allergies
- sensitivity to cold
- insomnia
- hiccups (cramp-relieving influence on diaphragm)
- fatigue
- migraines
- mastitis

Birch polypore growing on a tree in the authors' mushroom garden in Austria

Photo: Moritz Wildenauer

Photo: Benedikt Wurth

9. Recipes and Processing Edible Mushrooms

All recipes are for two to three people.

Shiitake

Shiitake is hugely popular in the kitchen. It is delicious and can be prepared in many ways. You can easily sauté it in oil or coat it in breadcrumbs, and it makes for tasty mushroom sauces. The mushrooms can be frozen, though we recommend preparing them beforehand by sautéing in oil and seasoning. Preserved in vinegar and herbs, they are a delicacy.

Depending on how they will be prepared, mushrooms can be harvested in different stages of development. Whatever the case, the cap should ideally be curved somewhat inward. In general, the stem is not used, though it can be dried and ground into powder. Dried, shiitake mushrooms store much longer.

Mushrooms can be harvested when still young, especially when they will be pickled or otherwise preserved. Do not try to harvest mushrooms at their maximum size, as the older they get the more likely they are to contain worms.

Often one sees small black beetles crawling around the gills. This is no cause for concern—insects can be removed during the harvest by lightly knocking on the mushroom cap. It is not necessary to wash the mushrooms.

Breaded Shiitake Mushrooms with Herb Sauce

Ingredients
- 15 medium to small shiitakes without stems
- 1 handful whole wheat or spelt flour
- 2 eggs (seasoned with herbal salt), beaten
- 1 handful bread crumbs
- 2 cups (½ L) coconut oil or lard for frying

For the sauce
- 1 cup (240 ml) sour cream
- 1 cup (240 ml) crème fraîche
- fresh or dried herbs
- salt, pepper
- 1 tsp curry, paprika, and other seasonings

Dredge shiitake mushrooms through the flour on both sides, dunk into beaten and seasoned eggs, and roll through bread crumbs (may need to be pressed on depending on their quality).

Heat a sufficient quantity of oil (should be ¾ in [1 to 2 cm] deep) in an appropriate pan with medium heat. Lay breaded mushrooms in heated oil and brown. Afterward let drip and, if desired, soak up excess oil on paper towels.

For the sauce, mix sour cream and crème fraîche with herbs and seasonings and serve with mushrooms as dip.

OUR TIP: *Breaded shiitakes freeze well; they maintain their flavor. When ready to eat, do not thaw them. Fry them as-is, otherwise they will become mushy.*

Stir-Fried Vegetables with Dried Shiitake Mushrooms

Ingredients
- 0.9 oz (25 g) dried shiitakes
- 1 small chile pepper, cut into small slices
- a dish of tamari (soy sauce) and water for soaking
- 1 red bell pepper
- 3.5 oz (100 g) green cabbage
- 2 green onions
- ½ leek
- fresh seasonal herbs (e.g., cilantro, lemon, basil, etc.)
- about ½ in (1 cm) ginger (or as desired)
- a bit of olive oil, coconut oil, butter, or lard
- salt, pepper
- 1 dish sprouts (e.g., lentil sprouts, mung bean sprouts)
- 1 tablespoon black sesame

Soak dried, thickly cut shiitakes with a few chile slices in tamari and water for about 20 minutes. Wash vegetables (bell pepper, cabbage, spring onions, leek) and cut to your preferred size. Finely cut fresh herbs and ginger (separately). Place vegetables and mushrooms with liquid in a heated wok or in a large pan with oil and sauté well. Later add the ginger. Stew everything together for about 5 minutes on medium heat. The vegetables can stay somewhat crunchy.

Salt and pepper to taste. Garnish with sprouts, fresh herbs, and sesame.

Rice noodles, stir-fried noodles, or basmati rice make for great accompaniments.

OUR TIP: *Sprouts are easily made. Soak the dried seeds of lentils or mung beans overnight. Then, spread evenly in a germination dish (they should not be too dense) and rinse twice a day with cold water. After two to three days—depending upon the temperature in the room—the sprouts will be ready and can be eaten raw with stir-fried vegetables or salad.*

Stir-fried vegetables with dried shiitake mushrooms

Shiitake spread or pesto

Garden Salad with Sautéed Shiitake Mushrooms

Ingredients
- seasonal greens (endive, lettuce, corn salad, arugula, etc.)
- several cocktail tomatoes
- 1 cucumber
- 1 red pepper
- 1 handful fresh shiitake mushrooms
- butter or oil for sautéing
- salt, pepper
- 1 packet (4 oz [(125 g]) mozzarella

For the dressing
- fresh seasonal herbs
- olive oil
- balsamic vinegar
- salt, pepper
- 1 tablespoon sour cream
- 1 teaspoon hot mustard
- lemon juice

Wash greens, cocktail tomatoes, cucumber, and pepper, cut small, and set aside.

For the dressing, chop fresh herbs and mix well with rest of ingredients.

Cut shiitake mushrooms into thin slices, sauté well in butter or oil, and season lightly. Cut mozzarella and put in salad.

Pour dressing over salad and serve on plates. Distribute shiitake among the salad plates.

DECORATION TIP: *If marigolds or nasturtiums happen to be in bloom, decorate the salad with them. They'll make for a colorful accent.*

Shiitake Spread or Pesto

Ingredients
- 1 lb (500 g) fresh shiitake mushroom caps
- 6 tablespoons olive oil, plus more to taste
- 2.8 oz (80 g) almonds
- 5 tablespoons Parmesan cheese
- 1 bunch parsley
- 1 dash lemon juice, plus more to taste
- salt
- ground pepper

Cut shiitake caps into slices and sauté well with oil in a pan. Roast almonds whole without oil, then chop finely. Grate Parmesan as finely as possible and mix with chopped almonds.

Cut up parsley and, along with a squirt of lemon juice, add to almond mixture.

Once shiitakes are browned, pulse in a blender. Add to almond mixture and mix well. Add olive oil, salt, pepper, and lemon juice to taste. If the mixture is too dry, simply add more olive oil.

OUR TIP: *Preservation made easy—pour into jars (not quite to top), seal, and pasteurize in rapidly boiling water for one hour. Allow to slowly cool thereafter. Keep oil or any spread off rim and lid when filling to ensure proper sealing. When well pasteurized and stored in a cool, dark place, it will keep for one year.*

Pickled Shiitake Mushrooms

Ingredients
- about 2.2 lb (1 kg) shiitake mushrooms

For the marinade
- 1 pint (½ L) mild apple cider vinegar
- 1 pint (½ L) water
- mustard seed
- peppercorns
- salt

Combine all ingredients for the marinade, seasoning with salt to taste.

Keeping small mushrooms whole and cutting down larger ones, add mushrooms to marinade and bring to a boil. Mushrooms should be completely covered by marinade. Allow to cool and, if necessary, season further.

Fill jars with mushrooms and marinade and seal well. Pasteurize jars for one hour in rapidly boiling water and allow to cool slowly.

Oyster Mushrooms

Oyster mushrooms differ not only in appearance—each type has its own characteristic flavor. The mushrooms are excellent for sauces and are great when simply seared and eaten on buttered bread.

Oyster mushrooms grow in clusters, which means that entire clusters should be harvested at once. If you attempt to only harvest the largest mushrooms, the smaller ones will no longer grow well and may die.

The stem is not used in cooking but it is still best to harvest the stem with the cluster so there are no leftover parts on the growing medium. If there is no soil or growing medium on the mushrooms, there is no need to wash them.

Oyster mushrooms can be stored in the refrigerator for up to four days at 36 to 39° F (2 to 4° C), but will probably grow more mycelium. This is normal and does not affect mushroom quality.

Oyster Mushrooms in Herb Sauce

Ingredients
- 7 oz (200 g) oyster mushrooms (whichever type has ripe fruit bodies)
- butter or cooking oil
- 2 onions or 1 leek, chopped
- 1 bunch fresh herbs (e.g., basil, parsley, cilantro)
- ½ cup (120 ml) heavy whipping cream
- salt, pepper
- seasonal vegetables (e.g., ½ zucchini, 1 bulb fennel, 2 carrots, 6 leaves Swiss chard, ¼ cabbage), thinly sliced
- seasoning (oregano, marjoram, salt, pepper, curry powder, garlic)
- 1 handful grated Parmesan cheese

Cut oyster mushrooms and sauté in oil with chopped onions or leek. Add sliced herbs and finally douse with cream. Add salt and pepper and season to taste.

Preserved shiitake mushrooms

In another pan, brown vegetables in oil and season well. Serve cooked vegetables and oyster mushrooms together with Parmesan cheese with a side dish of your choice.

OUR TIP: *Rice, quinoa, buckwheat, and fettucine are all appropriate sides.*

Filled Crepes with Oyster Mushrooms

Ingredients
- 14 oz (400 g) oyster mushrooms (whichever type has ripe fruit bodies)
- 1 onion
- 1 to 2 cloves garlic
- 1 tablespoon butter
- seasonal vegetables (carrots, eggplant, or zucchini), grated or thinly sliced
- dried herbs (e.g., thyme, basil, rosemary, etc.)
- 5¼ oz (150 g) mung bean sprouts
- salt, pepper
- 4 oz (100 g) gruyère/Swiss cheese or similar, grated
- butter for the dish

For the crepe batter
- 4 oz (100 g) flour
- 1 cup (¼ L) water
- 3 eggs
- salt
- frying oil (butter, coconut, lard)

Cut oyster mushrooms into small pieces and remove the hard stem. Cut onion and garlic as small as possible and heat with mushrooms and butter in a pan. Add grated carrots or other thinly sliced vegetables. Add chopped herbs and sprouts when done cooking. Salt and pepper to taste.

While cooking mushrooms and vegetables, heat oven to 300° F (180° C). For crepe batter—which could be prepared before the vegetables—mix flour, water, eggs, and salt into a batter and fry in a pan with butter or other oil.

Finally, fill crepes with vegetable-mushroom sauce, roll up, and place in a buttered casserole dish. Sprinkle with grated cheese and bake at 300° F (180° C) for 10 to 15 minutes.

OUR TIP: *A salad of arugula, lamb's lettuce (corn salad), and tomatoes goes well with this recipe.*

Grilled Oyster Mushrooms with Thyme

An outstanding dish in summer consists of grilled oyster mushrooms—no matter what type. Whether in the oven or on the grill, the fine aroma of these mushrooms unfolds and they become somewhat crispy.

Ingredients
- oyster mushrooms (blue oyster, golden oyster, late oyster, or pink oyster)
- melted butter or olive oil
- dried herbs (e.g., thyme, basil, rosemary, etc.)
- salt

Remove stems from mushrooms (wash dirty mushrooms, if needed), place in a bowl, and drizzle with melted butter or olive oil, though not too much. Season to taste with herbs and salt, spread on a baking sheet, and grill in convection oven preheated to 300° F (180° C) until lightly brown, which should take about 20 minutes. Mushrooms can also be laid on the grill on aluminum foil.

OUR TIP: *Add flair to this dish by serving it with an herb dip.*

Sheathed Woodtuft

Many mushroom connoisseurs include sheathed woodtuft in their list of tastiest mushrooms. It grows in clusters and its stems are fused at the base.

Only the caps are edible. Mushrooms often grow from the soil surrounding an inoculated log and may need to be washed.

Sheathed woodtuft mushrooms can be made into incredibly delicious soups and sauces or ground into powder.

Sheathed Woodtuft Soup

Ingredients
- 2 onions
- 1 clove garlic
- olive oil and butter
- 5 medium potatoes
- ½ lb (250 g) sheathed woodtuft or nameko mushrooms, chopped
- 3 cups (¾ L) vegetable broth
- 2 stalks lovage or parsley
- salt, pepper
- caraway
- seasonal herbs (e.g., thyme, basil, rosemary, etc.)
- 1 dash vinegar

Chop onions and garlic, then lightly sauté in olive oil and butter. Peel potatoes, dice, and add to onions along with chopped mushrooms. Cook all together briefly, then add vegetable broth and lovage.

Season to taste with salt, pepper, caraway, and herbs. When potatoes have softened, add a dash of vinegar and taste.

OUR TIP: *Do not use too much water in proportion to the quantity of mushrooms you are using, as the soup will taste best when thick.*

RECIPES AND PROCESSING EDIBLE MUSHROOMS

Photo: Benedikt Wurth

Button Mushrooms (Champignons)

Button mushrooms are excellent in sauces or seared in salads. The stem and base can be eaten just like the cap.

White button mushrooms stored in the refrigerator often turn brown and should be used within three to four days. Harvest mushrooms when the veil on the cap has just broken. Reminder: Twist button mushrooms away from their growing medium.

Couscous Salad with Button Mushrooms

Ingredients
- 7 oz (200 g) button mushrooms
- 2 green onions
- butter
- 1 cup (195 g) couscous (or other grain, such as millet)
- 2 tomatoes
- ½ packet sheep's milk cheese
- ½ avocado

For the dressing
- olive oil
- balsamic vinegar
- salt, pepper
- 1 teaspoon hot mustard
- 1 dash lemon juice

Thinly slice mushrooms and green onions, and sauté in a pan with butter. Cook couscous with twice as much water, then let cool. Thinly slice tomatoes, cheese, and avocado and combine with cooled couscous. For the dressing, combine all ingredients and pour over the salad. Add browned mushrooms and green onions.

OUR TIP: *A great, quick meal for hot summer days.*

Light Summer Salad with Farro and Button Mushrooms

Ingredients
- 1 cup (200 g) farro
- 6 button mushrooms
- butter
- 5 cherry tomatoes
- 1 green onion
- 1 red pepper
- 4.4 oz (125 g) halloumi (Cypriot goat's cheese), thinly sliced
- 1 small bunch parsley, finely chopped

For the dressing
- olive oil
- vinegar
- 1 dash lemon juice
- herb salt, pepper
- mustard
- ½ teaspoon honey

Cook farro until soft, then rinse until cold. Cut mushrooms into rough pieces, sauté in a pan with butter, and set aside. In the same pan, sear vegetables and cheese.

Serve mushrooms, vegetables, and cheese with cooled farro.

Combine all ingredients. Dress the salad and garnish with parsley.

King Oyster Mushrooms

This mushroom has a wonderful aroma and is much enjoyed in the kitchen. Caps and stems are both useable. Commercial king oyster mushrooms often have a "pot-bellied" stem, due to temperature regulation and humidity levels during growth. Thanks to its firm consistency, it goes well with vegetable dishes, but it is also good breaded or simply sautéed and served atop buttered bread with fresh herbs.

Chickpea and Tomato Stew with King Oyster Mushrooms

Ingredients
- 3 king oyster mushrooms
- 2 carrots
- ½ leek
- 1 large clove garlic
- 1 small chile pepper (optional)
- olive oil
- ½ cup (1/8 L) white wine
- 4.4 oz (125 g) chickpeas, soaked overnight
- 1 stalk finely chopped lovage
- 8.5 oz (240 g) can whole, skinned tomatoes
- 1 teaspoon honey or maple syrup
- 1 teaspoon lemon juice
- salt
- finely chopped parsley as garnish
- toasted bread for serving

Cut mushrooms, carrots, leek, garlic, and chile pepper finely, lightly sauté in oil, and douse with white wine.

Then add chickpeas, lovage, and skinned tomatoes and simmer on low heat. Add honey or maple syrup, lemon juice, and salt to taste. Serve with parsley and toasted bread.

Photo: Benedikt Wurth

Photo: Benedikt Wurth

Lion's Mane

In Asian cuisine, this mushroom is a true delicacy. With its incomparable forest aroma and its wild-looking fruit body, this mushroom deserves a place in every kitchen.

These tender, white mushrooms should be processed as soon after harvest as possible. Cut away carefully from its growing medium with a knife and, as much as possible, avoid bruising. There is no need to wash lion's mane.

Sautéed Lion's Mane in Vegetables with Quinoa and Cheese Patties

Ingredients
- 4 lion's mane mushrooms
- 2 small onions, finely sliced
- butter
- fresh herbs (e.g., ½ bunch lemon basil), finely chopped
- scant ½ cup (100 ml) heavy whipping cream
- edible flowers (e.g., nasturtium)

For the vegetable accompaniment
- 1 handful pointed white or savoy cabbage
- ¼ zucchini
- 3 carrots
- oil
- some white wine
- salt, pepper

For the quinoa and cheese patties
- ½ cup (85 g) quinoa
- ½ cup (100 g) rice
- 1 egg
- salt
- 1 teaspoon curry powder or turmeric
- sheep or goat's milk cheese, grated

Cook ½ cup quinoa and ½ cup rice in twice as much water. Pre-heat convection oven to 300° F (180° C).

In the meantime, cut vegetables into fine slices or dice, and grate the carrots. Heat oil in a pan, sauté vegetables, douse with white wine, and simmer on low heat. Season with salt and pepper to taste.

Combine cooked grains with egg, salt, and curry powder or turmeric. Form patties, sprinkle with grated cheese, and slide in the oven. Bake until patties have a golden yellow color and have become slightly crispy, then turn off heat and leave in warm oven.

Cut mushrooms into thin slices and sauté together with onion in butter in a pan. Then add herbs and douse with cream. Serve on warmed plates with edible flowers.

Wood Ear (Jelly Ear)

Wood ear has a fairly mild flavor and is used more for its consistency to enhance all sorts of dishes. It thickens soups well when dried and milled into powder. The mushroom has long been used in traditional Chinese medicine for its positive effects on the body.

Asian Soup with Wood Ear and Chile

Ingredients
- 5 oz (150 g) glass noodles (Chinese vermicelli)
- 2 chicken breasts
- 4 tablespoons (30 g) cornstarch
- oil, for sautéing
- 3 carrots (or pak choi)
- 1 red pepper
- 2 oz (50 g) cauliflower
- 5 oz (150 g) fresh wood ear mushrooms
- ½ in (1 cm) piece ginger
- 2 cloves garlic
- 1 quart (1 L) vegetable broth
- ½ bunch lemon verbena or lovage
- lemon juice

- tamari (or soy sauce)
- pepper
- 2 finely cut green onions for serving
- 2 medium-hot sliced chile peppers

For seasoning
- 1 dash tamari (or soy sauce)
- 1 teaspoon dashi (Japanese special seasoning)
- scant ¼ cup (50 ml) heavy whipping cream

Cook glass noodles in water according to instructions on package.

Cut chicken breasts into bite-sized pieces and dredge in cornstarch. To do this, put cornstarch in a sealable bag, then put in chicken, seal the bag, and shake lightly—cornstarch will be more evenly distributed this way.

Heat oil in a wok and briefly sauté chicken on all sides. Set aside in a bowl.

For the seasoning, mix tamari, dashi, and cream in a bowl and pour over chicken (1 cup [200 ml] liquid should suffice). Set aside for now; wait until later to add to soup.

Cut vegetables and mushrooms into bite-sized pieces and briefly sauté in the same wok. Cut ginger and garlic finely and add once vegetables and mushrooms have nearly finished cooking. Then, douse with vegetable broth and simmer all together for a while. Add lemon verbena or lovage to the broth to add a note of freshness, then remove before serving. Add chicken and drained noodles at the very end and briefly cook everything together. You may season additionally with a dash of lemon juice, tamari, and/or pepper.

Serve with finely cut green onions and freshly cut chile peppers.

King Stropharia (Garden Giant)

This firm-fleshed mushroom makes for delicious sauces and is great for breading. Its cultivation is well-established in Europe, and with little effort, you can grow this slightly nutty-tasting mushroom in your own garden.

King Stropharia Mushroom Sauce with Bread Dumplings

Ingredients
- About 5 cups (400 g) king stropharia mushrooms
- 1 large onion, chopped
- 2 carrots, grated
- ¼ celeriac, finely chopped
- herbs (e.g., thyme, basil, sage)
- scant ¼ cup (100 ml) heavy whipping cream
- salt, pepper
- freshly grated Parmesan cheese for serving

For the dumplings
- 1 onion
- 3.5 oz (100 g) butter
- about ½ lb (250 g) day-old rolls
- 3 eggs (separate yolks from whites)
- about 1 cup (¼ L) lukewarm milk
- ½ bunch parsley, chopped
- salt
- nutmeg, ground

For the dumplings, chop onion and lightly sauté in butter. Dice the day-old rolls and combine well with egg yolks, lukewarm milk, chopped parsley, salt, nutmeg, and sautéed onions. Allow to rest for about 25 minutes. Beat egg whites until stiff peaks form and combine with the day-old roll mixture.

Form a log and distribute this "dough" onto cloth napkins and close with strings. Bring water

Dried shiitake mushrooms

and a bit of salt to a boil in a sufficiently large pot. Place the dumpling (wrapped in napkins) in the water for 25 to 30 minutes, then remove and refresh with cold water.

While the dumpling cooks, slice mushrooms and sauté with onion, grated carrots, and diced celeriac in a pan. Once these ingredients are sufficiently cooked, add freshly cut herbs and douse with cream. Add salt and pepper to taste.

Cut dumpling into ½ in (1 cm) slices and serve together with mushroom sauce. Freshly grated Parmesan sprinkled on top provides the finishing touch.

Making Mushroom Powder and Processing Medicinal Mushrooms

Mushrooms are not only good for eating, they are also known for their healing effects. Mushrooms have been used for millennia as medicine. In previous chapters, we wrote of the cultivation and use of edible and medicinal mushrooms. Now we would like to focus on a practical aspect, the production of mushroom powder and mushroom tea.

The table opposite lists mushrooms that can either be cultivated at home or gathered in the wild. For information on other mushrooms with health benefits, see "The Use of Mushrooms in Medicine" (p. 108). For sources, see the appendix.

Drying Shiitake Mushrooms

When drying shiitake mushrooms, it is important to first cut them up into thin strips or to dice them. Lay the pieces on a baking sheet (with parchment paper) and dry in a convection oven at 122° F (50° C). Prop the oven door open slightly so that moisture can escape—a

Processing Methods for Home-Grown or Wild-Gathered Mushrooms

Mushroom	Cultivation Style	Processing Options
Shiitake (Lentinula edodes)	Inoculation of logs (beech, oak, birch)	edible tea powder dried
Sheathed woodtuft (Kuehneromyces mutabilis)	Inoculation of logs (birch, beech)	edible seasoning
Oyster mushroom (Pleurotus)	Logs (hardwood), straw	edible
Reishi (Ganoderma lucidum)	wood substrate inoculation of beechwood logs	tea extract powder
Almond mushroom (Agaricus blazei Murrill)	Compost	edible powder dried
Lion's mane (Hericium erinaceus)	growing kit inoculation of logs (birch, oak)	edible powder extract
Wood ear (Auricularia)	inoculation of logs (elder) gather wild (beech, elder)	edible powder
Birch polypore (Piptoporus betulinus)	Gather wild (only on birch)	powder extract tea

cooking spoon should do the trick. Otherwise, a food dehydrator will also work well.

When shiitake mushrooms are fully dry, they are breakable. Store in a sealable glass jar so they do not lose their aroma. When stored poorly, mushrooms take on moisture and quality suffers.

If you would like to cook some of your dried shiitakes, soak in hot water for five minutes beforehand. Or you can soak dried pieces of shiitake in tamari and water for 15 to 20 minutes, then cook the whole concoction.

OUR TIP: *It doesn't hurt to cover outdoor cultures during long rainy periods, as overly wet mushrooms do not dry well.*

Making Mushroom Salt

Because it is so tough, the stem of shiitake mushrooms is not used to make food. However, it can be dried and milled together with salt. This mushroom-salt mixture can be used to refine many dishes and is very healthy. Dried stems should be cut small so they will grind well in the mill.

Almond mushrooms and sheathed woodtuft are also good to use as seasoning in sauces or as mushroom salt.

Processing Reishi into Powder

Reishi has a very firm fruit body. With a sharp knife, you can cut it into thin slices, then reduce to smaller pieces. Like shiitake, it should be dried in a convection oven at a temperature of about

Reishi tea

Birch polypore in various stages of processing

122° F (50° C) or in a food dehydrator. Make sure everything dries completely, as otherwise mold may form in storage, rendering your reishi unusable. Sealable glass jars are the best storage containers for dried mushrooms at home.

To make tea out of your dried reishi pieces, grind them in a blender or with a spice mill. The finer you mill the reishi, the better the tea.

One heaped teaspoon of reishi powder per cup of tea is an ideal ratio. Enjoy on its own or as a mixture with other teas.

Mushroom Tea from Dried Birch Polypore Mushrooms

If you are gathering birch polypore from the wild, please confirm its identity with a good mushroom identification book (or, better, with multiple books). Like reishi, birch polypore is also good as a tea. Cut freshly gathered mushrooms into thin slices and dry in a convection oven at 104–122° F (40–50° C) or in a food dehydrator.

Store in well-sealed jars and mill as needed with a coffee or spice mill.

For tea, pour hot water over 2 teaspoons of powder and sieve after 15 minutes of steeping. If it is bitter, either mix with herbal tea or steep for a shorter length of time.

Freshly harvested shiitake mushrooms

10. Marketing Organic Mushrooms on a Small Scale

Once you have become experienced enough with cultivating mushrooms, you may wish to start a small business selling fresh mushrooms. We are not talking here about commercial cultivation on any grand scale, but rather the production of organic mushrooms for supplemental income or to accompany an assortment of vegetables you may already be marketing.

The demand for regional, high-quality products at farmers' markets, in Community Supported Agriculture (CSA) shares, or from restaurants is high. You can start a mushroom cultivating operation with a fairly minimal capital input and modest investment of time and energy. For farmers and market gardeners, mushrooms represent a relatively easy way to complement the assortment of products they are already bringing to market on a regular basis.

Case Study: Home Produced Fresh Mushrooms at Gasthaus Seidl, Vienna, Austria

Mr. Seidl presents an interesting model for the marketing of mushrooms. He runs a restaurant in Vienna's 3rd district. At the time of this writing, he has been growing edible mushrooms in the cellar under the restaurant for about a year and a half. He not only uses the best resources he can for growing media, but he also offers his guests a wide selection of edible mushrooms, such as king oyster mushrooms, pioppino, oyster mushrooms, and shiitake. Sustainable, regionally produced foods are valued more and more. Mr. Seidl's example shows how simple it can be to realize this ideal.

Elements of a Successful Organic Mushroom Business

Site	A space suitable for growing mushrooms must be available (shady area in garden, clean basement or cellar, etc.)
Raw materials (wood, straw, growing kit)	Organic raw materials readily available nearby
Experience	The mushroom being grown should be adapted to the environment and several successful crops should have been harvested before attempting to go professional (that is, start small, grow the operation as your experience grows)
Mushroom types	Mushrooms should be selected that are in demand from your clientele
Contacts	Enter into cooperative relationships with other mushroom growers (e.g., for making sterile substrate or procuring preinoculated growing media)
Storage	Refrigerator and/or cool room (such as root cellar) for storage
Sales	Farmers' markets, CSAs, restaurants, direct to customers

The cellar, which is two floors below the restaurant, has been retrofitted with a ventilation system and a sprinkler system to optimize growing conditions. Mr. Seidl is happy about the expansion of the menu, as customers cannot get enough of the home-grown mushrooms. The restaurant was once famous for its wine selection; now it is probably more famous for its mushrooms. If they are not able to use up all they grow, they sell the rest to mushroom connoisseurs in their neighborhood. In winter, they pickle, dry, and freeze mushrooms to sell in the restaurant.

Mr. Seidl is impressed with the freshness and quality of the mushrooms they are able to grow; commercially available mushrooms simply do not compare. They grow from substrates pre-inoculated by local mushroom growers, which helps keep the workload to a minimum. The main tasks are monitoring, harvesting, and keeping the area sanitary to ensure unwanted germs are kept away.

The in-house mushroom production has been quite the hit in Vienna, earning Mr. Seidl the nickname "Schwammerlwirt" (mushroom innkeeper).

Cultivating Mushrooms in the Context of Permaculture

Permaculture is about living in a way that provides for a better future and, in this context, producing some or all of your own food. As space is often at a premium, here is a list of possibilities for working around this:

- on a windowsill or on a balcony/patio: grow herbs, lettuce, and vegetables in pots or other containers
- participate in or establish a community garden
- shop at or start a food co-op or buying club for regional, organically produced foods
- buy into a CSA in which shareholders receive regular deliveries of farm products from the farm or group of farms selling shares

Essentially, it's about bringing food production back to the people and away from international conglomerates. The concepts of regionalism and quality become more tangible when you begin to grow and preserve your own food

MARKETING ORGANIC MUSHROOMS ON A SMALL SCALE

Oyster mushrooms as harbinger of spring

and begin to cooperate with other growers and farmers in your area.

It seems, then, only natural to integrate mushrooms in such a system, what with the diversity of cultivation methods that are appropriate to a wide range of sites.

In urban settings, you can grow mushrooms in a basement, cellar, backyard, shady balcony, or patio or in self-made growing media. Abandoned farm buildings can be adapted for mushroom cultivation. Another important issue is the sustainable utilization of raw materials. If, for example, you are removing an alder from the garden, the trunk (and, if large enough, its branches) along with the stump can be inoculated with mushroom spawn. Though the wood will no longer be good as a source of fuel, it will be integrated into a new natural cycle. The resulting partially decomposed wood can then be composted.

Practicing permaculture also means using minimal means to produce the best possible results. In this way, "garbage" becomes a raw material and incomplete cycles return to being closed loops.

OUR TIP: *For more on permaculture, see the book* Permaculture: Principles and Pathways Beyond Sustainability *by David Holmgren or, for German speakers,* Die kleine Permakultur-Fibel *by Bernard Gruber (www.permakultur.biz).*

Appendixes

Sources for Mushroom Spawn, Inoculated Logs, Growing Kits, and Medicinal Mushrooms

USA	UK
• Mushroom spawn, growing media, dowels, equipment, and kits: Field & Forest Products—fieldforest.net; Fungi Perfecti—fungi.com; Mushroom Mountain—mushroommountain.com • Mushroom kits: Mushroom Adventures—mushroomadventures.com • Powders and extracts: Mushroom Harvest—mushroomharvest.com • Information: North American Mycological Association—namyco.org	• Mushroom spawn, growing media, dowels, equipment, and kits: Mushroombox—mushroombox.co.uk • Mushroom spawn and kits: Gourmet Woodland Mushrooms—gourmetmushrooms.co.uk • Inoculated logs: Rustic Mushroom Company—rusticmushrooms.co.uk • Kits: Merryhill Mushrooms—merryhill-mushrooms.co.uk
Austria	**Germany**
• Mushroom spawn and inoculated logs: Waldviertler Pilzgarten—pilzgarten.at • Extracts, growing media, and indoor mini-greenhouses: Tyroler Glückspilze—gluckspilze.com • Growing media: Pilz-Kultur—pilz-kultur.at	• Growing media, information, extracts: Hawlik Vitalpilze: pilzshop.de • Information and free consultation: Myko-Troph—heilenmitpilzen.de • Extracts: Heilpilzversand Berlin—heilpilze-berlin.de

Sources for Laboratory Materials

Topic	Utensils or Materials	Source
Propagation material	• grain spawn • plug spawn	• sylvaninc.com; sporetradingpost.com; pilzgarten.at
	• pure strain • spore prints • growing kits • special cultures—medicinal mushrooms	• sylvaninc.com; sporetradingpost.com; pilzkultur.at; gluckspilze.com; hawlik-vitalpilze.de
Growing media and additives	• organic corn • untreated hardwood dowels • beechwood shavings • malt extract agar • malt extract • gypsum • lime	• farm supply, hardware store, or grocery store • laboratory specialist retailer
Lab paraphernalia	• disinfectants • hand sanitizer • alcohol • scalpels and inoculation loops • alcohol lamps or blowlamps • autoclave bags • glass jars with filters • filter discs • glass petri dishes • disposable petri dishes • Erlenmeyer flasks • scales (lb/oz [kg/g]) • measuring cups • silicone hoses • gloveboxes	• pharmacy or chemist Many utensils and other accessories you may already have around the house or can be found at a hardware store
	• gauze	• pharmacy or chemist
	• Laminar Air Flow	• laboratory specialist retailer
	• heat sealers	• office supply shop
	• pressure cookers	• household accessories shop
	• autoclaves	• laboratory specialist retailer • used online

Bibliography and Recommended Reading

English

- Russell, S. (2014): *The Essential Guide to Cultivating Mushrooms*. North Adams: Storey Publishing
- Stamets, P. (2000): *Growing Gourmet and Medicinal Mushrooms*. Third edition, New York: Ten Speed Press
- Stamets, P.; Chilton, J. (1983): *The Mushroom Cultivator: A practical guide to growing mushrooms at home*. Washington: Agarikon Press

German

- Bertram, H. (2014): Birkenporling. www.passion-pilze-sammeln.com/birkenporling.html
- Engelbrecht, J. (2004): *Pilzanbau in Haus und Garten*. Stuttgart: Ulmer
- Gesellschaft für Vitalpilzkunde e.V. (Ed.) (2009): *Vitalpilze: Naturheilkraft mit Tradition—neu entdecken*. First edition, Augsburg: GfV.
- Haidvogl, W. (2014): Unsere Pilze. http://www.pilz-kultur.at/Die%20Seite/index.php/unsere-pilze
- Hawlik, W. (1985): *Waldpilzzucht für jedermann: erfolgsanleitung für den Anbau von Waldpilzen auf Stroh und Holz*. Eighth edition, Munich: Dr. Richter GmbH
- Hawlik, W.; B. (2014): Pilze züchten. http://www.pilzshop.de/pilze+zuechten/13/ac
- Krämer, N. (2014): Anleitung für den Anbau von Braunkappe und Austernpilz auf Stroh. http://www.shiitake.de/an_duebel/an_stroh/index.html
- Mushroom Research Centre GmbH (o.J.): Pilzzucht für jedermann: Eine Einführung in die Weld der Pilzzucht. Innsbruck: MRCA
- Schmaus, F. (2009): *Die Natur als Apotheke nutzen: Heilen mit Pilzen*. Third edition, Limeshain: MykoTroph
- Schmidt, W. (2009): *Anbau von Speisepilzen: Kulturverfahren für den Haupt—und Nebenerwerb*. Stuttgart: Ulmer
- Svrcek, M. (1983): *Dausien's großes Pilzbuch in Farbe*. Prague: Werner Dausien Verlag
- Tyroler Glückspilze (2014): Mykorrhiza Planzendünger. https://gluckspilze.com/Mykorrhiza-Pflanzenduenger
- Urban, A.; Pla, T. (2014): Trüffel selbst züchten. http://www.trueffelgarten.at/
- Wurth, M.; H. (2014): Waldviertler Pilzgarten. http://www.pilzgarten.at/

Private interviews and conversations with Franz Seidl, owner of Gasthaus Seidl, and Walter Haidvogl, chemist and mushroom grower, were important and useful, informing several chapters of this book.

Index

Page numbers in italic type refer to illustrations.

A

agar *10*, 12, 16, 98–100, *101*, 104–5
Agaricus
  bisporus see button mushroom
  blazei Murrill see almond mushroom
Agrocybe aegerita see poplar mushroom
air filter 98, *102*
alder 18, *49*
almond mushroom 59, 65, *65*, 113
  medicinal use 112–13, 131
antibiotic nutrients 98
Armillaria spp. see honey fungus
Atlas cedar 70
auger method 17, 24, *24*, *26*, 75–6, *76*, *77*, 81
Auricularia auricula-judae see wood ear
autoclave 97
autumn 31, 40, 41
average household needs 15

B

beech 18, *19*, 42, 45, 49, 53, 54, 93
birch 18, 42, 45, 49, 53, 78, 93
birch polypore 93, 115–16, *116*, 131, 132, *132*
Botrytis cinerea 12
bran 16
brown-gilled woodlover 48–9, *48*

buna shimeji 59, 65, *65*
burgundy mushroom *see* king stropharia
button mushroom 8, 58–9, 61–4, *61*, *62*, 64
  compost 62–3
  growing sites 64
  harvesting 62–3, 64
  inoculating 63
  recipes 125–6

C

cap 11
casing 62
champignon *see* button mushroom
chanterelle 8, 9, 67
Chinese medicine 12, 108
cold frames 35, *39*, 56, 91
colonizing growing medium 15, 25, 38–41, 57, 71
companion plants 32–3, 72, *72*, 74, 80, *80*, 81–3, *82*
compost as growing medium 8, 14, 17
  button mushrooms 62–4
  inoculating 12, 63
  mushroom beds 70–4
conifer tuft *see* brown-gilled woodlover
container-grown mushrooms 80–3, *80*, *82*, 87–8, *89*
Coprinus comatus see shaggy mane
cultivation overview *10*
cup fungus 94

D

decomposers 8, 12, 64

E

elder 52, 53, 78–9, 131
enokitake (enoki) 15, 18, 48, 50, *52*, 78
  harvesting 31
  mycorestoration 75

F

Flammulina velutipes see enokitake (enoki)
ferns 14
fields, cultivation in 67–77
fruit body *9*, 11
fruit trees 18, 21, 68
fungus gnats 21, 58, 91–2

G

Ganoderma lucidum see reishi
garden cultivation 11, 14–54, *14*, 26–8, 32–3
  colonization and storage 25, *25*, 38–41
  companion plants 32–3, 72, *72*, 74
  drainage 14
  irrigation 14, 28–30
  location 14
  moisture 14, 26
  mushroom beds 70–4
  shade 14, 26, 32–3, 35
garden giant *see* king stropharia
germination 11
gills 11
glovebox 97–8, *98*, 101

INDEX

greenhouses
   35, *84*, 85, 86–7, *87*, 94
   indoor 85, 86, *86*
Grifola frondosa 75
growing kits *10*
growing media 8, 15
   see also compost; straw; wood
   colonizing 15, 25, 38–41, 57, 71
gypsum 16, 63

H

harvesting
   garden mushrooms 30–1, 39, 41
   mushroom beds 74
   mushroom kits 61, 62–3
hen of the woods 75
Hericium erinaceus see lion's mane
honey fungus 8, 93
hornbeam 18, 42, 45, 54, 70
humus 8
hygiene 95, 96–8
hyphae 11
   mycorrhizae 9
   propagation 16–17, *16*
Hypholoma capnoides see brown-gilled woodlover
Hypsizygus
   tessulatus see buna shimeji
   ulmarius see oyster mushrooms (elm)

I

indoor cultivation 56–66, *58*, 92
   fermenting straw 56–7, 92
inky cap 94, *95*
inoculation *see individual species and growing media*

J

jelly ear *see* wood ear

K

kerf method
   17, 19–22, *20*, *21*, 25, 26, *26*, 27
king stropharia 15, 35, *36*, 37–9, 48, 50, *52*, 85, 91
   inoculating 15, 37–8, *37*
   mushroom beds 71, *71*, 74
   mycorestoration 75
   recipe 129–30
kits 58–64, 92
Kuehneromyces mutabilis see sheathed woodtuft

L

labeling 21, *21*, 79
larch 75
Lentinula edodes see shiitake
lifecycle 9, 11
linden 18
lion's mane 52, *52*, 53, 58, 59, 85, *88*, *96*, 111
   harvesting 31
   medicinal use 110–11, 131
   recipe 128
logs *see* wood as growing medium
luminescent panellus 33–4, *33*, 83

M

maitake 75
manure 63
maple 18, 45, 49, 53
marketing 133–5
medicinal mushrooms
   12, 108–16, 130–2
mice 70, 90

microorganisms, maceration by 8
mite infestations 92–3, *92*
moisture
   garden cultivation 14, 26
   mushroom kits 58, 62
   wood as growing medium 19, 25, 27
moss 14, 62, 93
molds 12, 17, 90, 94–5, 108
mushroom beds 70–4, 91
   harvesting 74
mycelium 9, 11, 18, 26
   cultivation *10*, 16–17, 16, 25, 102–3
   spreading to soil 26–7, *27*
   straw pellets 57, *57*
mycorestoration 74–5
mycorrhizae 8–9, 12, 67
   see also truffles
hyphae 9

N

nameko 31, 46, 48–9, *51*, 83
non-native species 68

O

oak
   18, 21, 42, 45, 49, 53, 70, 78–9
oyster mushrooms 8, *13*, 42, 56, 78–9, 91, *102*, *135*
   blue 30, 36, 41, 44, *44*, 45, 57, 78–9, 81, 83
   colonizing 41, 57
   container-grown *80*, 81, 83
   elm 41, 45, 46, 74
   golden 30, 36, 41, 42, 44, 45, 57, 74, 78–9, 81, 83, *88*, 90
   harvesting 31, 40, 41, 57–8
   indoor cultivation 56–8, *58*

oyster mushrooms (*continued*)
  Italian (lung) 36, 41, 44, 45, 47, 57, 74, 78, *80*, 83
  king 45, 46, 58–9, *60*, 126
  kits 58–9
  late *43*, 44, 45, 57, 78, 79
  medicinal use 115
  mushroom beds 71, 74
  mycorestoration 75
  outdoor cultivation 15, 18, 25, 26, 36, 40–1, 71
  pink 35, 41, *44*, 45, 46, 57, 85, *88*
  processing 131
  recipes 122–3, 126
  stump inoculation 75

P
Panellus stipticus 33–4, *33*
parasitic fungi 8, 12
penicillin 12, 108
pests and diseases 21, 27, 90–5, 97
  buried logs 28
  protected environments 85
pH value 98
Pholiota 75
  nameko see nameko
Phytophthora infestans 12
pine 70, 75
pinhead 9
pioppino see poplar mushroom
Piptoporus betulinus 115–16, *116*
Pleurotus see oyster mushrooms
plug method 16, 17, 22–3, *22*, *23*, *26*, 76, 77, 99–100, *99*, 105–6, *106*
polyculture 72, *72*, 74

polypore species 93
poplar 18, 53, 75, 78
poplar mushroom 59, 66, *66*
porcini 8, 9, 67
pores 11
portobello mushroom *64*
potato blight 12
pressure cooker 97
primordia 9, 11
propagation
  from tissue sample *10*, 11–12, 16, 102–3
  growing media 98–9, *99*
  hygiene 96–7
  laboratory 11, 96–106
  overview *10*, 11–12
  spawn 16–17, *16*
  spores 11, 16
protected environments 85–9

R
reishi 54, *55*, 58–9, 87–8, *87*, 89, *107*, *108*
  medicinal use 109, 131–2
  mycorestoration 75
rhizomorphic mycelium 9, 11
ridges 11
roof, protective 85–6, *85*

S
saplings, truffle 67, 70
saprophytic fungi 8
shade 14, 26, 32–3, 35, 71, 72, 80, 82, *82*, 83
shaggy mane 8, 71–4, *72*, *73*, *112*
  medicinal use 111–12
sheathed woodtuft 8, 15, 18, 19, 26, 46, 48, *48*, 49, 78–9, 83, 124
  container-grown 81

harvesting 30, 31
mycorestoration 75
processing 131
soup 124
shelf fungi 93
shiitake *14*, 42, *42*, 78, *110*, *133*
  container-grown 83
  drying 130–1
  garden cultivation 15, 18, 19, 25, 26, *26*, 42
  growing medium 42
  harvesting 30, 31, 42
  indoor cultivation 58
  kerf method 19, 21, 25, 26, *26*, 42
  kits 58–9, 61
  logs as fencing 34–5, *34*, *35*
  medicinal use 108, 110, 130–1
  mycelium 26, *26*
  positioning logs 28, *28*
  recipes 118–22
  watering logs 29–30, *29*, *30*
slime molds 93, *93*
slugs 15, 27, 28, 39, 42, 48, 56, 71–2, 90–1
spawn
  commercial 16–17
  cultivation *10*, 16–17, *16*
  grain 16–17, *16*, 21, 24, 37, 56–7, 70, 99, *99*, 100, 103–4, *104*, *106*
  inoculation see *individual growing media*
  mushroom beds 70–1
  propagation 96–106
  storage 17
split gill 93
spores 9, 11, 12, *102*
  positively and negatively

INDEX

polarized 11
propagation from 11, 16, 102–6
spore prints 102
spring 31, 40, 41
spruce 49, 75
stem 11
straw as growing medium 8, *10*, 14, 15, 17, 35–41, 91, 92, 94
   advantages 36
   colonization 38–41
   contact with soil 15
   fermentation 56–7, 92
   harvesting mushrooms 35, 39
   inoculating 12, 15, 37–8, *37*, 57
   king stropharia 15, 35, 37–9, *37*, 50, *85*
   mushroom beds 70–4, 91
   mushroom houses 87
   oyster mushrooms 15, 36, 42, 45
   positioning bales 35, 40
   preparation 37, 39, 40
   soil contact 35
   spent substrate 58
   straw pellets *10*, 45, 56–9, 57, *99*
   straw quality 36–7
   temperature 35
Stropharia rugoso-annulata see king stropharia
stump inoculation *10*, 24, *24*, 25, 75–6, *76*, 77, 78–9
sulphur tuft 18, 93
summer 31, 40
symbiosis 9, 67

T

teeth 11
Tilletia caries 12
tissue culture 102–3
Trüffelgarten 67, 68, 69
truffle fly 70
truffles 8, 67–70, *69*
Turkish filbert 70

W

watering
   garden cultivation 28–30
   mushroom beds 71
   mushroom houses 87
   mushroom kits 58–9
wheat, common bunt 12
willow 18, 49, 50, 53, *76*, 78
wine cap stropharia see king stropharia
wine production 12
winter 31, 40, 58–64, 72
wolf's-milk fungi 93, *93*
wood ear 52–3, *52*, 78–9, *113*, *114*
medicinal use 113, 115
processing 131
recipe 128–9
wood as growing medium 8, 14, 15, 17, 18–35, *91*, 94
   advantages 36
   auger method 17, 24, *24*, *26*, 81
   bark quality 18–19, 25
   buried logs 10, 15, 21, 26–8, *26*, *27*, 30, *32*, 81–3, *91*, 93
   colonization and storage 25, *25*, 26
   enokitake 15
   hanging logs *28*
   harvesting months 30–1
   inoculating 12, 15, 17, 19–23, 81, 82
   kerf method 17, 19–22, *20*, *21*, 25, 26, *26*, 27–8
   log size 19, 21, 30
   moisture content 19, 25, 27, 28
   moss covering 27, *27*, 93
   oyster mushrooms 15, 42, 45
   plug method 16, 17, 22–3, *22*, *23*, 26
   pre-inoculated 15, 81
   sheathed woodtuft 15
   shiitake see shiitake
   stacked logs *11*, *14*, 15, 21, 28
   storage 25, *25*, 26
   tree stumps see stump inoculation
   tree trunks and branches 10, 11
   watering logs 28–30
wood chip beds *10*, 15, 70–4, 93
wood quality 18
woodlands 12, 14, 67–77
woodwarts 93, *93*

Y

yeasts 12
yellow fieldcap 94
Yellow panels 58, 92

Acknowledgments

Our thanks go to Löwenzahn Verlag, who motivated us to write this handbook. They have made it possible to share our experience with all those who wish to test the waters of organic mushroom growing.

Thank you to all who have made this book possible: Anita Winkler and Petra Möderle for your capable and dedicated support and to graphic designer Judith Eberharter.

We would like to thank all those who have supported us throughout the journey of writing this book.

Thank you to our families and our friends.

In particular we thank Moritz Wildenauer, Benedikt Wurth, Rosa-Maria Binder, and Anna Folie for their numerous, wonderful photographs.

And finally we would like to thank wonderful Nature herself, for everything she gives us and for all she allows to thrive.

About the Translator

Ian Miller is the author of T*he Scything Handbook* (Filbert Press, 2017) and has translated several books on ecological topics including *The Manual of Seed Saving* and *Cultivating Chaos*. He lives in northeast Iowa with his wife and two daughters. He is on the web at www.iangmiller.net